S0-EUL-017

3 1611 00342 6894

TK 5611 .F54 1972

Field, Henry Martyn
History of the
Atlantic telegraph.

48092

UNIVERSITY LIBRARY
Governors State University
Park Forest South, Il. 60466

HISTORY

OF THE

ATLANTIC TELEGRAPH

BED OF THE ATLANTIC, NORTH AND SOUTH, THROUGH THE CAPE DE VERDS, AZORES, AND TELEGRAPH PLATEAU.

HISTORY

OF THE

ATLANTIC TELEGRAPH

BY

HENRY M. FIELD

'Tis not in mortals to command success;
But we'll do more, Sempronius—we'll deserve it."
ADDISON'S CATO.

BOOKS FOR LIBRARIES PRESS
FREEPORT, NEW YORK

First Published 1866
Reprinted 1972

INTERNATIONAL STANDARD BOOK NUMBER:
0-8369-6768-2

LIBRARY OF CONGRESS CATALOG CARD NUMBER:
76-38351

PRINTED IN THE UNITED STATES OF AMERICA
BY
NEW WORLD BOOK MANUFACTURING CO., INC.
HALLANDALE, FLORIDA 33009

PREFACE.

MANKIND worship success but think too little of the means by which it is attained. What toil and patience have gone before; what days and nights of watching and weariness; how often hope deferred has made the heart sick; how year after year has dragged on, and seen the end still afar off—all that counts for little, if the long struggle do not close in victory. And yet in the history of human achievements, it is necessary to trace these beginnings step by step, if we would learn the lesson they teach, that it is only out of heroic patience and perseverance that any thing truly great is born.

The object of this volume is to record the history of a great enterprise, which, after many disappointments, seems at last to have touched its hour of triumph. It is a work which has cost its projector twelve years of constant toil, and more than fifty voyages, of which two thirds have been across the Atlantic, and the rest to Newfoundland; and which has been pursued in the face of a thousand difficulties, and, what was harder still, of a public incredulity, which sneered at every failure, and derided the attempt as a delusion and a dream. Against such discouragements nothing could avail but that faith, or fanaticism, which, believing the incredible, achieves the impossible. The story of such an enterprise deserves to be told. The relation of the writer to the principal actor in this work, has given him peculiar facilities for obtaining information on all points necessary to an authentic history; but he trusts it will not lead him to overstep the strictest limits of modesty. His object is not to exalt an individual, but to give a faithful record, that shall bear in every line the stamp of truth; and to do justice to *all*, on both sides the Atlantic, who have borne a part in a work which will do so much to link together two great nations, and to promote the peaceful intercourse of mankind.

NEW-YORK, August, 1866.

CONTENTS.

CHAPTER I.

DISCOVERY OF THE NEW WORLD, BY COLUMBUS. RELATIVE POSITION OF THE TWO HEMISPHERES. NEAREST POINTS—THE OUTLYING ISLANDS, IRELAND AND NEWFOUNDLAND. FIRST SUGGESTION OF A SHORTER WAY TO EUROPE. LETTER OF BISHOP MULLOCK. THE ELECTRIC TELEGRAPH COMPANY OF NEWFOUNDLAND. LABORS OF MR. F. N. GISBORNE. FAILURE OF THE COMPANY.

CHAPTER II.

MR. GISBORNE COMES TO NEW-YORK. IS INTRODUCED TO CYRUS W. FIELD, WHO CONCEIVES THE IDEA OF A TELEGRAPH ACROSS THE ATLANTIC OCEAN. IS IT PRACTICABLE? TWO ELEMENTS TO BE MASTERED, THE SEA AND THE LIGHTNING. INQUIRIES ADDRESSED TO LIEUTENANT MAURY AND PROFESSOR MORSE. ANSWER OF LIEUTENANT MAURY. VISIT OF PROFESSOR MORSE. MR. FIELD DETERMINES TO EMBARK IN THE UNDERTAKING.

CHAPTER III.

EFFORTS TO ENGAGE CAPITALISTS IN THE ENTERPRISE. PETER COOPER, MOSES TAYLOR, MARSHALL O. ROBERTS, AND CHANDLER WHITE. COMMISSION SENT TO NEWFOUNDLAND. THEY OBTAIN A NEW CHARTER FOR THE NEW-YORK, NEWFOUNDLAND, AND LONDON TELEGRAPH COMPANY. RETURN TO NEW-YORK. THE CHARTER IS ACCEPTED, THE COMPANY ORGANIZED, AND THE CAPITAL RAISED.

CHAPTER IV.

THE LAND-LINE BEGUN IN NEWFOUNDLAND. IMMENSE UNDERTAKING. FOUR HUNDRED MILES OF ROAD TO BE BUILT. TWO YEARS OF LABOR. FIRST ATTEMPT TO LAY A CABLE ACROSS THE GULF OF ST. LAWRENCE, IN 1855. FAILURE. SECOND ATTEMPT, IN 1856, WHICH IS SUCCESSFUL.

CHAPTER V.

The Deep-Sea Soundings. The Old Method of Ball and Line. Massey's Indicator. Invention of Lieutenant Brooke. Cruise of the Dolphin in 1853, and of the Arctic in 1856. The brave Lieutenant Berryman. Soundings by Commander Dayman, of the British Navy, in the Cyclops, in 1857. The Bed of the Atlantic. Depths in Different Parts. The Telegraphic Plateau. Submarine Mountain off the Coast of Ireland.

CHAPTER VI.

Mr. Field goes to England to organize the Atlantic Telegraph Company. Confers with John W. Brett. Seeks Counsel of Engineers and Electricians. Result of Experiments. Applies to the Government for Aid. Letter from the Treasury. Enters into an Agreement with Messrs. Brett, Bright, and Whitehouse to form a Company. The Enterprise brought before the British Public. Capital raised and Company organized. Choice of a Board of Directors. Contract for the Cable.

CHAPTER VII.

Mr. Field returns to America. Starts immediately for Newfoundland. Returns and goes to Washington, to seek Aid from the American Government. Opposition in Congress. The Cable among the Politicians. Debate in the Senate. Support of Mr. Seward and Mr. Rusk. Bill passed.

CHAPTER VIII.

Return to England. The Niagara—Captain Hudson. The Agamemnon. Expedition of 1857. Sailing from Ireland. Speech of the Earl of Carlisle. The Cable broken.

CHAPTER IX.

Preparations for a Second Expedition. Mr. Field is made the General Manager of the Company. Mr. Everett and the Paying-out Machine. The Valorous takes the Place of the Susquehanna. The Squadron assemble at Plymouth. They go

TO SEA, JUNE 10. HEAVY GALE. THE AGAMEMNON IN DANGER OF BEING FOUNDERED. THE CABLE LOST THREE TIMES. THE SHIPS RETURN TO ENGLAND. MEETING OF THE DIRECTORS. SHALL THEY ABANDON THE PROJECT? ONE MORE TRIAL.

CHAPTER X.

THE SHIPS SAIL ON A SECOND EXPEDITION. THEY MEET IN MID-OCEAN. SUCCESSFUL VOYAGE. CABLE LANDED IN IRELAND AND NEWFOUNDLAND.

CHAPTER XI.

NEWS OF THE SUCCESS. GREAT EXCITEMENT IN AMERICA. POPULAR ENTHUSIASM. CELEBRATION IN NEW-YORK AND OTHER CITIES.

CHAPTER XII.

SUDDEN STOPPAGE OF THE CABLE. REACTION OF PUBLIC FEELING. GRAVE SUSPICIONS OF BAD FAITH. DID THE CABLE EVER WORK? DECISIVE PROOF.

CHAPTER XIII.

ATTEMPTS TO REVIVE THE COMPANY. PROFOUND DISCOURAGEMEET. IT APPLIES TO THE GOVERNMENT FOR AID, WHICH DECLINES TO GIVE AN UNCONDITIONAL GUARANTEE. FAILURE OF THE RED SEA TELEGRAPH. SCIENTIFIC EXPERIMENTS. CABLES LAID IN THE MEDITERRANEAN AND THE PERSIAN GULF. EFFORTS TO RAISE CAPITAL IN THE UNITED STATES AND IN ENGLAND. BRIEF HISTORY OF THE NEXT FIVE YEARS.

CHAPTER XIV.

THE ENTERPRISE RENEWED. PROPOSALS FOR ANOTHER CABLE. GREAT IMPROVEMENT ON THE OLD. GENEROUS OFFER OF THE MANUFACTURERS TO TAKE HALF THE CAPITAL. THE WORK BEGUN. THE GREAT EASTERN AND CAPTAIN ANDERSON. THE WHOLE CABLE SHIPPED ON BOARD IN THE SPRING OF 1865. EXPEDITION IN JULY.

CONTENTS.

CHAPTER XV.

RESULT OF THE EXPEDITION OF 1865. CONFIDENCE STRONGER THAN EVER. INSTANT RESOLVE TO RAISE THE BROKEN END OF THE CABLE, TO COMPLETE IT TO NEWFOUNDLAND, AND TO LAY ANOTHER LINE. NEW SHARES ISSUED. METHOD DECLARED UNAUTHORIZED BY LAW. FORMATION OF THE ANGLO-AMERICAN TELEGRAPH COMPANY. CAPITAL RAISED. NEW CABLE MADE AND SHIPPED ON BOARD THE GREAT EASTERN.

CHAPTER XVI.

THE EXPEDITION OF 1866. VICTORY AT LAST.

HISTORY OF THE ATLANTIC TELEGRAPH.

CHAPTER I.

DISCOVERY OF THE NEW WORLD, BY COLUMBUS. RELATIVE POSITION OF THE TWO HEMISPHERES. NEAREST POINTS—THE OUTLYING ISLANDS, IRELAND AND NEWFOUNDLAND. FIRST SUGGESTION OF A SHORTER WAY TO EUROPE. LETTER OF BISHOP MULLOCK. THE ELECTRIC TELEGRAPH COMPANY OF NEWFOUNDLAND. LABORS OF MR. F. N. GISBORNE. FAILURE OF THE COMPANY.

WHEN Columbus sailed from the shores of Spain, it was not in search of a New World, but only to find a nearer path to the Old. He sought a western passage to India. He had come to the belief that the earth was round; but he did not once dream of another continent than the three which had been the ancient abodes of the human race—Europe, Asia, and Africa. All the rest was "the great deep." Hence he believed that he could sail direct from Spain to India; that over that space, covering, as he then supposed, one third of the round globe, the billows rolled without a shore. No undiscovered continent existed even in his imagination. Nay, after he had crossed the Atlantic, and descried the green woods of San Salvador rising out of the Caribbean Sea, he thought he saw before him one of the islands of the Asiatic coast. Cuba he believed was a part of the mainland of India; Hayti was the Ophir of King Solomon; and when, on a later voyage, he came to the broad mouth of the Orinoco, and saw it pouring its mighty flood into the Atlantic, he rejoiced that he had found the great river Gihon, which had its rise in the garden of Eden! Even to the hour of his death, he remained ignorant of the real extent of his magnificent discovery. It was reserved to later times to lift the curtain fully from the world of waters; to reveal the true magnitude of the globe; and to unite the distant hemispheres by ties such as the great discoverer never knew.

It is hard to imagine the darkness and the terror which then hung over the face of the deep. The ocean to the west was a Mare Tenebrosum—a Sea of Darkness, into which the boldest voyagers feared to venture. Columbus was the most daring navigator of his time. He had made voyages to the Western Islands, to Madeira and the Canaries, to Iceland on the north, and to the Portuguese settlements in Africa. But when he came to cross the sea, he had to grope his way almost blindly. But a few rays of knowledge glimmered, like stars, on the pathless waters. Hence, when he sailed on his voyage of discovery, he directed his course, not west, but *south*. A chart, made by the eminent Italian geographer Toscanelli, represented the eastern coast of Asia as lying opposite to the western coast of Africa. Hence Columbus first sailed along the latter as far as the Canary Islands, and thence bore away for India!

From this route taken by the great navigator, he crossed the ocean in its widest part. Had he, instead, followed the track of the Northmen, who crept around from Iceland to Greenland and Labrador; or had he sailed straight to the Azores, and then borne away to the north-west, he would much sooner have descried land from the mast-head. But steering in darkness, he crossed the Atlantic where it is broadest *and deepest;* where, as submarine explorers have since shown, it rolls over mountains, lofty as the Alps and the Himmalehs, which lie buried beneath the surface of the deep. But farther north the two continents, so widely sundered, incline toward each other, till the bold headlands of Newfoundland stand over against those of Ireland, even as the white chalk cliffs of England gleam across the Channel from France.

As the island of NEWFOUNDLAND is to stand in the foreground of this history, it is in place here to speak of its geographical position and its importance. It holds the same relation to America that Ireland does to Europe. Stretching far out into the Atlantic, it is the vanguard of the western continent, or rather the signal-tower from which the New World may speak to the Old.

Nor is it without other claims to importance, which ought to be recognized. In extent, it is equal to England. Is it not surprising that an island large enough for a kingdom, lying off our own coast, should be so little known? And yet the reason is obvious. It lies out of the track of European commerce. Our ships, though they skirt the Banks of Newfoundland, pass a few leagues to the south, and get only a distant glimpse of its rocky shores. Even what is seen gives the country. rather an ill reputation. It has a rock-bound coast, around which hang perpetual fogs and mists, through which great icebergs, breaking from the Northern Sea, drift slowly down, like huge phantoms of the deep, gliding away to be dissolved by the warm breath of the Gulf Stream. The remembrance of these chilling fogs and threatening icebergs makes the voyager shiver as he recalls that dangerous coast.

Sailing west from Cape Race, and making the circuit of the island as far as the Straits of Belle Isle, one is often reminded of the most northern peninsula of Europe. The rocky shores are indented with numerous bays, reaching far up into the land, like the fiords along the coast of Norway; while the large herds of Caribou deer, that are seen feeding on the hills, might easily be mistaken for the flocks of reindeer that browse on the pastures and drink of the mountain torrents of ancient Scandinavia.

The interior of the island is little known. It is uninhabited and almost unexplored. It is a boundless waste of rock and moor, where vast forests stretch out their unbroken solitudes, and the wild bird utters its lonely cry. Bears and wolves roam on the mountains. Especially common is the large and fierce black wolf; while of the smaller animals, whose skins furnish material for the fur-trade, such as martins and foxes, there is the greatest abundance. But from all pests of the serpent tribe, Newfoundland is as free as Ireland, which was delivered by the prayers of St. Patrick. There is not a snake or a frog or a toad in the island!

Yet, even in this ruggedness of nature, there is a wild beauty, which only needs to be "clothed upon" by the hand of man. Newfoundland, in many of its features, is not unlike Scotland, even in its most desolate portions, where vast fields of rock, covered with thick moss, remind the emigrant Scot of the heather on his native moors. In the interior are lakes as long as Loch Lomond, and mountains as lofty as Ben Lomond and Ben Nevis. There are passes as wild as the vale of Glencoe, which make one feel that he is in the heart of the Highlands, while the roar of the torrents yet more vividly recalls the

> Land of the brown heath and shaggy wood,
> Land of the mountain and the flood.

Yet in all this there is nothing to repel human habitation. By the hand of industry, these wild moors might be transformed into fruitful fields. We think it a very cold country, where winter reigns over half the year, as in Greenland; yet it is not so far north as Scotland, nor is its climate more inhospitable. It only needs the same population, the same hardy toil; and the same verdure would creep up its hill-sides, which now makes green and beautiful the loneliest of Scottish glens.

But at present the country is a *terra incognita*. In the interior, there are no towns, and no roads. As yet, almost the whole wealth of the island is drawn from the sea. Its chief trade is its fisheries, and the only places of importance are a few small towns, chiefly on the eastern side, which have grown up around the trading posts. Besides these, the only settlements are the fishermen's huts scattered along the coast. Hence the bishop of the island, when he would make his annual visit to his scattered flock, is obliged to sail around his diocese in his private yacht, since even on horseback it would not be possible to make his way through the dense forests to the remote parts of the island. Indeed, it was this circumstance that first suggested the idea of cutting across the island a nearer way, not only for the people themselves, but for communication between Europe and America.

It was in one of these excursions around the coast that the good Bishop Mullock, the head of the Roman Catholic Church in Newfoundland, when visiting the western portion of his diocese, lying one day becalmed in his yacht, in sight of Cape North, the extreme point of the province of Cape Breton, bethought himself how his poor neglected island might be benefited by being taken into the track of communication between Europe and America. He saw how nature had provided an easy approach to the mainland on the west. About sixty miles from Cape Ray stretched the long island of Cape Breton, while, as a stepping-stone, the little island of St. Paul's lay between. So much did it weigh upon his mind that, as soon as he got back to St. John's, he wrote a letter to one of the papers on the subject. As this is the first suggestion that I have found of a telegraph across Newfoundland, I here give his letter in full:

To the Editor of the Courier:

Sir: I regret to find that, in every plan for transatlantic communication, Halifax is always mentioned, and the natural capabilities of Newfoundland entirely overlooked. This has been deeply impressed on my mind by the communication I read in your paper of Saturday last, regarding telegraphic communication between England and Ireland, in which it is said that the nearest telegraphic station on the American side is Halifax, twenty-one hundred and fifty-five miles from the west of Ireland. Now would it not be well to call the attention of England and America to the extraordinary capabilities of St. John's, as the nearest telegraphic point? It is an Atlantic port, lying, I may say, in the track of the ocean steamers, and by establishing it as the American telegraphic station, news could be communicated to the whole American continent forty-eight hours, *at least*, sooner than by any other route. But how will this be accomplished? Just look at the map of Newfoundland and Cape Breton. From St. John's to Cape Ray there is no difficulty in establishing a line passing near Holy-Rood along the neck of land connecting Trinity and Placentia Bays, and thence in a direction due west to the Cape. You have then about forty-one to forty-five miles of sea to St. Paul's Island, with deep soundings of one hundred fathoms, so that the electric cable will be perfectly safe from icebergs. Thence to Cape North, in Cape Breton, is little more than twelve miles. Thus it is not only practicable to bring America two days nearer to Europe by this route, but should the telegraphic communication between England and Ireland, sixty-two miles, be realized, it presents not the least difficulty. Of course, we in Newfoundland will have nothing to do with the erection, working, and maintenance of the telegraph; but I suppose our Government will give every facility to the company, either English or American, who will undertake it, as it will be an incalculable advantage to this country. I hope the day is not far distant when St. John's will be the first link in the electric chain which will unite the Old World and the New. J. T. M.

St. John's, November 8, 1850.

This suggestion proved to be seed sown on good ground, since out of it in a great measure sprang the first attempt to link the Island of Newfoundland with the mainland of America. For about the same time, the attention of Mr. Frederick N. Gisborne, a telegraph operator, was attracted to a similar project. Being a man of great quickness of mind, he instantly saw the importance of such a work, and took hold of it with enthusiasm. It might easily occur to him without suggestion from any source. He had had much experience in telegraphs, and was then engaged in constructing a telegraph line in Nova-Scotia. Whether, therefore, the idea was first with him or with the bishop, is of little consequence. It might occur at the same time to two intelligent minds, studying the public good, and be alike honorable to both.

But having taken hold of this idea, Mr. Gisborne pursued it with indomitable resolution. As the labors of this gentleman were most important in the beginning of this work, I am happy to bear the fullest testimony to his zeal and energy. For this purpose, I quote from a letter written by Mr. E. M. Archibald, now British Consul at New-York, and formerly Attorney-General of Newfoundland:

"It was during the winter of 1849-50, that Mr. Gisborne, who had been, as an engineer, engaged in extending the electric telegraph through Lower Canada and New-Brunswick to Halifax, Nova Scotia, conceived the project of a telegraph to connect St. John's, the most easterly port of America, with the main continent. The importance of the geographical position of Newfoundland, in the event of a telegraph ever being carried across the Atlantic, was about the same time promulgated by Dr. Mullock, the Roman Catholic Bishop of Newfoundland, in a St. John's newspaper.

"In the spring of the following year, (1851,) Mr. Gisborne visited Newfoundland, appeared before the Legislature, then in session, and explained the details of his plan, which was an overland line from St. John's to Cape Ray, nearly four hundred miles in length, and (the submarine cable between Dover and Calais not having then been laid) a communication between Cape Ray and Cape Breton by steamer and carrier-pigeons, eventually, it was hoped, by a submarine cable across the Gulf of St. Lawrence. The Legislature encouraged the project, granted £500 sterling to enable Mr. Gisborne to make an exploratory survey of the proposed line to Cape Ray, and passed an act authorizing its construction, with certain privileges, and the appointment of commissioners for the purpose of carrying it out. Upon this, Mr. Gisborne, who was then the chief officer of the Nova Scotia Telegraph Company, returned to that province, resigned his situation, and devoted himself to the project of the Newfoundland telegraph. Having organized a local company for the purpose of constructing the first telegraph line in the island, from St. John's to Carbonear, a distance of sixty miles, he, on the fourth of September, set out upon the arduous expedition of a survey of the proposed line to Cape Ray, which occupied upward of three months, during which time himself and his party suffered severe privations, and narrowly escaped starvation, having to traverse the most rugged and hitherto unexplored part of the island.* On his return, hav-

* "On the fourth day of December, I accomplished the survey through three hundred and fifty miles of wood and wilderness. It was an ar-

ing reported to the Legislature favorably of the project, and furnished estimates of the cost, he determined to proceed to New-York, to obtain assistance to carry it out. . . . Mr. Gisborne returned to St. John's in the spring of 1852, when, at his instance, an act, incorporating himself (his being the only name mentioned in it) and such others as might become shareholders in a company, to be called the Newfoundland Electric Telegraph Company, was passed, granting an exclusive right to erect telegraphs in Newfoundland for thirty years, with certain concessions of land, by way of encouragement, to be granted upon the completion of the telegraph from St. John's to Cape Ray. Mr. Gisborne then returned to New-York, where he organized, under this charter, a company, of which Mr. Tebbets and Mr. Holbrook* were prominent members, made his financial arrangements with them, and proceeded to England to contract for the cable from Cape Ray to Prince Edward Island, and from thence to the mainland. Returning in the autumn, he proceeded in a small steamer, in the month of November of that year, 1852, to stretch the first submarine cable, of any length, in

duous undertaking. My original party, consisting of six white men, were exchanged for four Indians; of the latter party, two deserted, one died a few days after my return, and the other, 'Joe Paul,' has ever since proclaimed himself an ailing man."—*Letter of Mr. Gisborne.*

* Horace B. Tebbets and Darius B. Holbrook.

America, across the Northumberland Strait from Prince Edward Island to New-Brunswick, which cable, however, was shortly afterward broken, and a new one was subsequently laid down by the New-York, Newfoundland, and London Telegraph Company. In the spring of the following year, 1853, Mr. Gisborne set vigorously to work to complete his favorite project of the line (which he intended should be chiefly underground) from St. John's to Cape Ray. He had constructed some thirty or forty miles of road, and was proceeding with every prospect of success, when, most unexpectedly, those of the company who were to furnish the needful funds dishonored his bills, and brought his operations to a sudden termination. He and the creditors of the company were for several months buoyed up with promises of forthcoming means from his New-York allies, which promises were finally entirely unfulfilled; and Gisborne, being the only ostensible party, was sued and prosecuted on all sides, stripped of his whole property, and himself arrested to answer the claims of the creditors of the company. He cheerfully and honorably gave up every thing he possessed, and did his utmost to relieve the severe distress in which the poor laborers on the line had been involved."

This is a testimony most honorable to the engineer who first led the way through a pathless wilderness But this Newfoundland scheme is not to be confound ed with that of the Atlantic Telegraph, which did not come into existence until a year or two later. The latter was not at all included in the former. Indeed, Mr. Gisborne himself says, in a letter referring to his original project: "My plans were to run a subterranean line from Cape Race to Cape Ray, fly carrier-pigeons and run boats across the Straits of Northumberland to Cape Breton, and thence by overland lines convey the news to New-York." Though he adds : " Meanwhile Mr. Brett's experimental cable between Dover and Calais having proved successful, I set forth in my report, [which appeared a year after his first proposal,] that 'carrier-pigeons and boats would be required only until such time as the experiments then making in England with submarine cables should warrant a similar attempt between Cape Ray and Cape Breton.'" But nowhere in his report does he allude to the possibility of ever thus spanning the mighty gulf of the Atlantic.

But several years after, when the temporary success of the Atlantic Telegraph gave a name to every body connected with it, he or his friends seemed not unwilling to have it supposed that this was embraced in the original scheme. When asked why he did not publish his grand idea to the world, he answers: "Because I was looked upon as a wild visionary by my friends, and pronounced a fool by my relatives for re-

America, across the Northumberland Strait from Prince Edward Island to New-Brunswick, which cable, however, was shortly afterward broken, and a new one was subsequently laid down by the New-York, Newfoundland, and London Telegraph Company. In the spring of the following year, 1853, Mr. Gisborne set vigorously to work to complete his favorite project of the line (which he intended should be chiefly underground) from St. John's to Cape Ray. He had constructed some thirty or forty miles of road, and was proceeding with every prospect of success, when, most unexpectedly, those of the company who were to furnish the needful funds dishonored his bills, and brought his operations to a sudden termination. He and the creditors of the company were for several months buoyed up with promises of forthcoming means from his New-York allies, which promises were finally entirely unfulfilled; and Gisborne, being the only ostensible party, was sued and prosecuted on all sides, stripped of his whole property, and himself arrested to answer the claims of the creditors of the company. He cheerfully and honorably gave up every thing he possessed, and did his utmost to relieve the severe distress in which the poor laborers on the line had been involved."

This is a testimony most honorable to the engineer who first led the way through a pathless wilderness But this Newfoundland scheme is not to be confound ed with that of the Atlantic Telegraph, which did not come into existence until a year or two later. The latter was not at all included in the former. Indeed, Mr. Gisborne himself says, in a letter referring to his original project: "My plans were to run a subterranean line from Cape Race to Cape Ray, fly carrier-pigeons and run boats across the Straits of Northumberland to Cape Breton, and thence by overland lines convey the news to New-York." Though he adds : " Meanwhile Mr. Brett's experimental cable between Dover and Calais having proved successful, I set forth in my report, [which appeared a year after his first proposal,] that 'carrier-pigeons and boats would be required only until such time as the experiments then making in England with submarine cables should warrant a similar attempt between Cape Ray and Cape Breton.'" But nowhere in his report does he allude to the possibility of ever thus spanning the mighty gulf of the Atlantic.

But several years after, when the temporary success of the Atlantic Telegraph gave a name to every body connected with it, he or his friends seemed not unwilling to have it supposed that this was embraced in the original scheme. When asked why he did not publish his grand idea to the world, he answers : " Because I was looked upon as a wild visionary by my friends, and pronounced a fool by my relatives for resigning a lucrative government appointment in favor of such a laborious speculation as the Newfoundland connection. Now had I coupled it at that time with an Atlantic line, all confidence in the prior undertaking would have been destroyed, and my object defeated." This may have been a reason for not announcing such a project to the public, but certainly it was not a reason for not imparting the secret confidentially to his friends. A man can hardly lay claim to that which he holds in such absolute reserve.

However, whether he ever entertained the *idea* of such a project, is not a matter of the slightest consequence to the public, nor even to his own reputation. Probably hundreds had a vague notion that such a thing might come to pass at some future day, just as many believe that flying-machines will yet navigate the air. *Ten years before*, Professor Morse had expressed, not a dreamer's fancy, but a deliberate conviction, founded on scientific experiments, that " a telegraphic communication might with certainty be established across the Atlantic Ocean;" so that the idea was not original with Mr. Gisborne, nor with others who have seemed anxious to claim its paternity.

It is a curious part of the history of great enterprises, that the moment one *succeeds*, a host spring up to claim the honor. Thus when, in 1858, the Atlantic Telegraph seemed to be a success, the public, knowing

24   HISTORY OF THE ATLANTIC TELEGRAPH.

well who had borne the brunt and burden of the undertaking, awarded him the praise which he so well deserved; but instantly there were other Richmonds in the field. Those who had had no part in the labor, at least claimed to have originated the idea! Of course, these many claims destroy each other. But after all, to raise such a point at all is the merest trifling. The question is not who first had the "idea," but who took hold of the enterprise as a practical thing; who grappled with the gigantic difficulties of the undertaking, and fought the battle through to victory?

As to Mr. Gisborne, his activity in the beginning of the Newfoundland telegraph is a matter of history. In that preliminary work, he bore an honorable part, and acquired a title to respect, of which he cannot be deprived. All honor to him for his enterprise, his courage, and his perseverance!

But for the company of which he was the father, which he had got up with so much toil, it lived but a few months, when it became involved in debt some fifty thousand dollars, chiefly to laborers on the line, and ended its existence by an ignominious failure. The concern was bankrupt, and it was plain that, if the work was not to be finally abandoned, it must be taken up by stronger hands.

CHAPTER II.

MR. GISBORNE COMES TO NEW-YORK. IS INTRODUCED TO CYRUS W. FIELD, WHO CONCEIVES THE IDEA OF A TELEGRAPH ACROSS THE ATLANTIC OCEAN. IS IT PRACTICABLE? TWO ELEMENTS TO BE MASTERED, THE SEA AND THE LIGHTNING. INQUIRIES ADDRESSED TO LIEUTENANT MAURY AND PROFESSOR MORSE. ANSWER OF LIEUTENANT MAURY. VISIT OF PROFESSOR MORSE. MR. FIELD DETERMINES TO EMBARK IN THE UNDERTAKING.

MR. GISBORNE left Halifax and came to New-York in January, 1854. Here he took counsel with his friend Tebbets and others; but they could give him no relief. It was while in this state of suspense that he met, at the Astor House, Mr. Matthew D. Field, an engineer who had been engaged in building railroads and suspension-bridges at the South and West. Mr. Field listened to his story with interest, and engaged to speak of it to his brother, Cyrus W. Field, a merchant of New-York, who had retired from business the year before, and had spent six months in travelling over the mountains of South-America, from which he had lately returned. Accordingly, he introduced the subject, but found his brother disinclined to embark in any new undertaking. Though still a young man, his life had been for many years one of incessant devotion to business. He had accumulated an ample fortune, and was not disposed to renew the cares, the anxieties, and the fatigues of his former life. But listening to the details of a scheme which had in it much to excite interest, and which by its very difficulty stimulated the spirit of enterprise, he at length consented to see Mr. Gisborne, and sent to invite him to his house. Accordingly he came, and spent an evening describing the route of his proposed telegraph, and the points it was to connect. After he left, Mr. Field took the globe which was standing in the library, and began to turn it over. *It was while thus studying the globe that the idea first occurred to him, that the telegraph might be carried further still, and be made to span the Atlantic Ocean.* This idea, as will soon appear, was not original with Mr. Field, though he was to be the instrument, in the hands of Providence, to carry it out. It was indeed a new idea *to him;* but it had long been a matter of speculation with scientific minds, though their theories had never attracted his attention. But once he had grasped the idea, it took strong hold of his imagination, and led him to entertain the Newfoundland scheme, as preliminary to the other. Had the former stood alone, he would never have undertaken it. He cared little about shortening communication with Europe merely by a day or two, by relays of boats and carrier-pigeons! But it was the hope of further and grander results that inspired him, and gave him courage to enter on a work of which no man could foresee the end.

But so vast an enterprise was not to be rashly undertaken. There were scientific problems involved, which could only be solved by scientific men, and perhaps not even by them; which, it might be, could only be answered by the final test of experiment. Before giving any definite reply to Gisborne, Mr. Field determined to apply to the highest authorities this side the Atlantic. The project of an Atlantic Telegraph involved two problems : Could a cable be stretched across the ocean ? and if it were, would it be good for any thing to convey messages ? The first was a question of mechanical difficulties, requiring a careful survey of the ocean itself, fathoming its depth, finding out the character of its bottom, whether level, or rough and volcanic ; and all the obstacles that might be found in the winds that agitate the surface above, or the mighty currents that sweep through the waters below. The second problem was one less mechanical, but more purely scientific, involving questions as to the laws of electricity, not then fully understood, and on which the boldest might feel that he was venturing on uncertain ground.

Such were the two elements or forces of nature to be encountered—the ocean and the lightning. Could they be controlled by any power of man? The very proposal was enough to stagger the faith even of an enthusiast. Who could lay a bridle on the neck of the wild sea? The attempt seemed as idle, if not as impious, as that of Xerxes to bind it with chains. Was it possible to combat the fierceness of the winds and waves, and to stretch one long line from continent to continent? And then, after the work was achieved, would the lightning run along the ocean-bed from shore to shore? Such were the questions which have puzzled many an anxious brain, and which now troubled the one who was to undertake the work.

To get some light in his perplexity, Mr. Field, the very next morning after his interview with Gisborne, wrote two letters, one to Lieutenant Maury, then at the head of the National Observatory at Washington, on the nautical difficulties of the undertaking, asking if the sea were itself a barrier too great to be overcome; and the other to Professor Morse, inquiring if it would be possible to telegraph over a distance so great as that from Europe to America?

The mail soon brought an answer from Lieutenant Maury, which began: "Singularly enough, just as

I received your letter, I was closing one to the Secretary of the Navy on the same subject." A copy of this he inclosed to Mr. Field, as giving his matured opinion. It has since been published. We give the greater part of it, to show the conclusions to which, even at that early day, scientific men were beginning to arrive:

"NATIONAL OBSERVATORY,
WASHINGTON, February 22, 1854.

"SIR: The United States brig Dolphin, Lieutenant Commanding O. H. Berryman, was employed last summer upon especial service connected with the researches that are carried on at this office concerning the winds and currents of the sea. Her observations were confined principally to that part of the ocean which the merchantmen, as they pass to and fro upon the business of trade between Europe and the United States, use as their great thoroughfare. Lieutenant Berryman availed himself of this opportunity to carry along also a line of deep-sea soundings, from the shores of Newfoundland to those of Ireland. The result is highly interesting, in so far as the bottom of the sea is concerned, upon the question of a submarine telegraph across the Atlantic; and I therefore beg leave to make it the subject of a special report.

"This line of deep-sea soundings seems to be decisive of the question as to the practicability of a submarine telegraph between the two continents, *in so far as the bottom of the deep sea is concerned.* From Newfoundland to Ireland, the distance between the nearest points is about sixteen hundred miles;* and the bottom of the sea between the two places is a plateau, which seems to have been placed there especially for the purpose of holding the wires of a submarine telegraph, and of keeping them out of harm's way. It is neither too deep nor too shallow; yet it is so deep that the wires but once landed, will remain for ever beyond the reach of vessels' anchors, icebergs, and drifts of any kind, and so shallow, that the wires may be readily lodged upon the bottom. The depth of this plateau is quite regular, gradually increasing from the shores of Newfoundland to the depth of from fifteen hundred to two thousand fathoms, as you approach the other side. The distance between Ireland and Cape St. Charles, or Cape St. Lewis, in Labrador, is somewhat less than the distance from any point of Ireland to the nearest point of Newfoundland. But whether it would be better to lead the wires from Newfoundland or Labrador is not now the question; nor do I pretend to consider the question as to the possibility of finding *a time*

---

* From Cape Freels, Newfoundland, to Erris Head, Ireland, the distance is sixteen hundred and eleven miles; from Cape Charles, or Cape St. Lewis, Labrador, to ditto, the distance is sixteen hundred and one miles.

calm enough, the sea smooth enough, a wire long enough, a ship big enough, to lay a coil of wire sixteen hundred miles in length; though I have no fear but that the enterprise and ingenuity of the age, whenever called on with these problems, will be ready with a satisfactory and practical solution of them.

"I simply address myself at this time to the question in so far as *the bottom of the sea* is concerned, and as far as that, the greatest practical difficulties will, I apprehend, be found after reaching soundings at either end of the line, and not in the deep sea. . . .

"A wire laid across from either of the above-named places on this side will pass to the north of the Grand Banks, and rest on that beautiful plateau to which I have alluded, and where the waters of the sea appear to be as quiet and as completely at rest as it is at the bottom of a mill-pond. It is proper that the reasons should be stated for the inference that there are no perceptible currents, and no abrading agents at work at the bottom of the sea upon this telegraphic plateau. I derive this inference from a study of a physical fact, which I little deemed, when I sought it, had any such bearings.

"Lieutenant Berryman brought up with Brooke's deep-sea sounding apparatus specimens of the bottom from this plateau. I sent them to Professor Bailey, of West-Point, for examination under his microscope.

This he kindly gave, and that eminent microscopist was quite as much surprised to find, as I was to learn, that all those specimens of deep-sea soundings are filled with microscopic shells; to use his own words, '*not a particle of sand or gravel exists in them.*' These little shells, therefore, suggest the fact that there are no currents at the bottom of the sea whence they came; that Brooke's lead found them where they were deposited in their burial-place after having lived and died on the surface, and by gradually sinking were lodged on the bottom. Had there been currents at the bottom, these would have swept and abraded and mingled up with these microscopic remains the *débris* of the bottom of the sea, such as ooze, sand, gravel, and other matter; but not a particle of sand or gravel was found among them. Hence the inference that these depths of the sea are not disturbed either by waves or currents. Consequently, a telegraphic wire once laid there, there it would remain, as completely beyond the reach of accident as it would be if buried in air-tight cases. Therefore, so far as the bottom of the deep sea between Newfoundland, or the North Cape, at the mouth of the St. Lawrence, and Ireland, is concerned, the practicability of a submarine telegraph across the Atlantic is proved. . . .

"In this view of the subject, and for the purpose of hastening the completion of such a line, I take the liberty of suggesting for your consideration the propriety of an offer from the proper source, of a prize to the company through whose telegraphic wire the first message shall be passed across the Atlantic.

"I have the honor to be, respectfully, etc.,
"M. F. MAURY,
"Lieutenant United States Navy.
"Hon. J. C. DOBBIN, Secretary of the Navy.

The reply of Professor Morse showed equal interest in the subject, in proof of which he wrote that he would come down to New-York to see Mr. Field about it. A few days after, he came, and saw Mr. Field at his house. This was the beginning of an acquaintance which soon ripened into friendship, and which henceforth united these gentlemen together in this great achievement. Professor Morse, in conversation, entered at length into the laws of electricity as applied to the business of telegraphing, and concluded by declaring his entire faith in the undertaking as a practicable thing; as one that might, could, and would be achieved. Indeed, this faith he had avowed years before. In a letter written as early as August tenth, 1843, to John C. Spencer, then Secretary of the Treasury, Professor Morse had detailed the results of certain experiments made in the harbor of New-York to show the power of electricity to communicate at great distances, at the close of which he says —in words that now seem prophetic : "*The practical inference from this law is, that a telegraphic communication on the electro-magnetic plan may with certainty be established across the Atlantic Ocean! Startling as this may now seem, I am confident the time will come when this project will be realized.*"

It was the good fortune of Mr. Field—at that time and ever since—to have at hand an adviser in whose judgment he had implicit confidence. This was his eldest brother, David Dudley Field. They lived side by side on Gramercy Park, and were in daily communication. To the prudent counsels, wise judgment, and unfaltering courage of the elder brother, the Atlantic Telegraph is more indebted than the world will ever know, for its first impulse and for the spirit which sustained it through long years of discouragement and disaster, when its friends were few. To this, his nearest and best counsellor, Mr. Field opened the project which had taken possession of his mind; and being strengthened by that maturer judgment, he finally resolved that, if he could get a sufficient number of capitalists to join him, he would embark in an enterprise which, beginning with the line to Newfoundland, involved in the end nothing less than an attempt to link this New World which Columbus had discovered, to that Old World which had been for ages the home of empire and of civilization. How the scheme advanced through the next twelve years, it will be our province to relate.

CHAPTER III.

EFFORTS TO ENGAGE CAPITALISTS IN THE ENTERPRISE. PETER COOPER, MOSES TAYLOR, MARSHALL O. ROBERTS, AND CHANDLER WHITE. COMMISSION SENT TO NEWFOUNDLAND. THEY OBTAIN A NEW CHARTER FOR THE NEW-YORK, NEWFOUNDLAND, AND LONDON TELEGRAPH COMPANY. RETURN TO NEW-YORK. THE CHARTER IS ACCEPTED, THE COMPANY ORGANIZED, AND THE CAPITAL RAISED.

AND so the young New-York merchant set out to carry a telegraph across the Atlantic Ocean! The design had in it at least the merit of audacity. But whether the end was to be sublime or ridiculous time alone could tell. Certain it is that when his sanguine temper and youthful blood stirred him up to take hold of such an enterprise, he little dreamed of what it would involve. He thought lightly of a few thousands risked in an uncertain venture; but never imagined that he might yet be drawn on to stake upon its success the whole fortune he had accumulated; that he was to sacrifice all the peace and quiet he had hoped to enjoy; and that for twelve years he was to be almost without a home, crossing and re-crossing the sea, urging his enterprise in Europe and America. But so it is, that the Being who designs great things for human welfare, and would accomplish them by human instruments, does not lift at once the curtain from the stern realities they are to meet, nor reveal the rugged ascents they are to climb; so that it is only when at last the heights are attained, and they look backward, that they realize through what they have passed.

But could he find any body to join him in his bold undertaking? Starving adventurers there always are, ready to embark in any Quixotic attempt, since they have nothing to lose. But would men of sense and of character; men who had fortunes to keep, and the habit which business gives of looking calmly and suspiciously at probabilities; be found to put capital in an enterprise where, if it failed, they would find their money literally at the bottom of the sea? It seemed doubtful, but he would try. His plan was, if possible, to enlist ten capitalists, all gentlemen of wealth, who together could lift a pretty heavy load; who, if need were, could easily raise a million of dollars, to carry out any undertaking.

The first man whom he addressed was Mr. Peter Cooper, who was then and is still his next-door neighbor. Here he found the indisposition which a man of large fortune—now well advanced in life—would naturally feel to embark in new enterprises. The reluctance in this case was not so much to the risking of capital, as to having his mind occupied with the care which it would impose. These objections slowly yielded to other considerations. As they talked it over, the large heart of Mr. Cooper began to see that, if it were *possible* to accomplish such a work, it would be a great *public* benefit. This consideration prevailed, and what would not have been undertaken as a private speculation, was yielded to public interest. The conference ended by a conditional agreement. Mr. Cooper would engage in it, if several others did. In the end, as we shall soon see, he became the President of the Company, and as such has remained to this day.

The early accession of this gentleman gave strength to the new enterprise. In all the million inhabitants of the city of New-York there is not a name which is better known, or more justly held in honor, than that of Peter Cooper. A native of the city, where he has passed his whole life, he has seen its growth, from the small town it was just after the Revolution, and has himself grown with it. Beginning with very small means and limited opportunities, he has become one of the great capitalists of the New World. But many who thus rise to wealth, in the process of accumulation, form penurious habits which cling to them, and to the end of their days it is the chief object of life to hoard and to keep. But

Mr. Cooper, while acquiring the fortune, has had also the heart, of a prince; and has used his wealth with a noble generosity. In the centre of New-York stands to-day a massive building, consecrated "To Science and Art," on which he has already expended six hundred and fifty thousand dollars, and which he has made a free gift to the city. His object was to benefit the poor but respectable young men and women of New-York. Remembering his own limited advantages of education, he desired that the young men of New-York, the apprentices and mechanics, should have better opportunities than he had ever enjoyed. For this he endowed courses of lectures on the natural sciences; he opened the largest reading-room in America, which furnishes a pleasant resort to a thousand readers daily; while to help the other sex, he added a School of Design for Women, which sends forth hundreds well fitted to be teachers, and some of them artists; and who go forth into the world to earn an honest living, and to bless the memory of their generous benefactor. This noble institution, standing right in the heart of New-York, will remain, long after its founder has passed away, his enduring monument.

Yet while doing so much for the public, those who see him in his family know how he retains the simple habits of early life—how, while giving hundreds of thousands to others, he cares to spend little on himself;

how he remains the same modest, kindly old man; the pure, the generous, and good. The accession of such a man to the head of the new Company was a benediction. It brought a blessing with it; and if in future years a hundred cables should link the New World to the Old, it will be a pleasant remembrance that the beginning of the enterprise was connected with that honored name.

Mr. Field next addressed himself to Mr. Moses Taylor, a well-known capitalist of New-York, engaged in extensive business reaching to different parts of the world, and whose daily observation of all sorts of enterprises, both sound and visionary, made him perhaps a severer judge of any new scheme. With this gentleman he had then no personal acquaintance, but sent a note of introduction from a friend, with a line requesting an interview, to which Mr. Taylor replied by an invitation to his house on an evening when he should be disengaged. As these two gentlemen have since been very intimately associated, they recur pleasantly to their first interview. Says Mr. Field: "I shall never forget how Mr. Taylor received me. He fixed on me his keen eye, as if he would look through me: and then, sitting down, he listened to me for nearly an hour without saying a word." This was rather an ominous beginning. However, his quick mind soon saw the possibilities of the enterprise, and the evening ended by an agreement—conditional, like Mr. Cooper's—to enter into it.

Mr. Taylor, being thus enlisted, brought in his friend, Mr. Marshall O. Roberts—a man whose career has been too remarkable to be passed without notice. A native of the city of New-York, (though his father was a physician from Wales, who came to this country early in this century,) he found himself, when a boy of eight years, an orphan, without a friend in the world. From that time he made his way purely by his own industry and indomitable will. At the age of twenty he was embarked in business for himself, and his history soon became a succession of great enterprises. If we were to relate some of the incidents connected with his rise of fortune, they would sound more like romance than reality. He was the first to project those floating palaces which now ply the waters of the Hudson and the great lakes. He was one of the early promoters of the Erie Railroad. When the discovery of gold in California turned the tide of emigration to that coast, he started the line of steamers running to the Isthmus of Panama, and controlled largely the commerce with the Pacific. Thus his hand was felt, giving impulse to many different enterprises on land and sea. His whole course has been marked by a spirit of commercial daring, which men call rashness, until they see its success, and then applaud as marvellous sagacity.

Mr. Field next wrote to Mr. Chandler White, a personal friend of many years standing, who had retired from business, and was living a few miles below the city, near Fort Hamilton, at one of those beautiful points of view which command the whole harbor of New-York. He too was very slow to yield to argument or persuasion. Why should he—when he had cast anchor in this peaceful spot—again embark in the cares of business, and, worst of all, in an enterprise the scene of which was far distant, and the results very uncertain? But enthusiasm is always magnetic, and the glowing descriptions of his persuader at length prevailed.*

There were now five gentlemen enlisted; and Mr. Field was about to apply to others, to make up his proposed number, when Mr. Cooper came to ask why *five* would not do as well as *ten ?* The question was no sooner asked than answered. To this all agreed, and at once fixed an evening when they should meet at Mr. Field's house to hear his statements and to ex-

* Although it is anticipating a year in time, I cannot resist the pleasure of adding here the name of another eminent merchant, who afterward joined this little Company, Mr. Wilson G. Hunt. Mr. Hunt is one of the old merchants of New-York who, through his whole career, has maintained the highest reputation for commercial integrity, and whose fortune is the reward of a long life of honorable industry. He joined the Company in 1855, and has been a strong and steady friend from that day to this.

amine the charter of the old company, find out what it had done, and what it proposed to do, what property it had and what debts it owed; and decide whether the enterprise offered sufficient inducements to embark in it. Accordingly they met, and for four nights in succession discussed the subject. It was in the dining-room of Mr. Field's house, and the large table was spread with maps of the route to be traversed by the line of telegraph, and with plans and estimates of the work to be done, the cost of doing it, and the return which they might hope in the end to realize for their labor and their capital. The result was an agreement on the part of all to enter on the undertaking, if the Government of Newfoundland would grant a new charter conceding more favorable terms. To secure this it was important to send at once a commission to Newfoundland. Neither Mr. Cooper, Mr. Taylor, nor Mr. Roberts would go; and it devolved on Mr. Field to make the first voyage on this business, as it did the more than fifty voyages since, either to Newfoundland, or across the Atlantic. But not wishing to take the whole responsibility, he was accompanied at his earnest request by Mr. White, and by Mr. D. D. Field, whose counsel, as he was to be the legal adviser of the Company, was all-important in the framing of the new charter that was to secure the rights of the Company. Says the latter gentleman, in an account given afterwards :

"The agreement with the Electric Telegraph Company, and the formal surrender of its charter, were signed on the tenth of March, [1854,] and on the fourteenth we left New-York, accompanied by Mr. Gisborne. The next morning we took the steamer at Boston for Halifax, and thence, on the night of the eighteenth, departed in the little steamer Merlin for St. John's, Newfoundland. Three more disagreeable days, voyagers scarcely ever passed, than we spent in that smallest of steamers. It seemed as if all the storms of winter had been reserved for the first month of spring. A frost-bound coast, an icy sea, rain, hail, snow and tempest, were the greetings of the telegraph adventurers in their first movement towards Europe. In the darkest night, through which no man could see the ship's length, with snow filling the air and flying into the eyes of the sailors, with ice in the water, and a heavy sea rolling and moaning about us, the captain felt his way around Cape Race with his lead, as the blind man feels his way with his staff, but as confidently and as safely as if the sky had been clear and the sea calm; and the light of morning dawned upon deck and mast and spar, coated with glittering ice, but floating securely between the mountains which form the gates of the harbor of St. John's. In that busy and hospitable town, the first person to whom we were introduced was Mr. Edward M. Archibald, then Attorney-General of the Colony, and now British Consul in New-York. He entered warmly into our views, and from that day to this, has been an efficient and consistent supporter of the undertaking. By him we were introduced to the Governor, (Kerr Bailey Hamilton,) who also took an earnest interest in our plans. He convoked the Council to receive us, and hear an explanation of our views and wishes. In a few hours after the conference, the answer of the Governor and Council was received, consenting to recommend to the Assembly a guarantee of the interest of £50,000 of bonds, an immediate grant of fifty square miles of land, a further grant to the same extent on the completion of the telegraph across the ocean, and a payment of £5000 toward the construction of a bridle-path across the island, along the line of the land telegraph."

This was a hopeful beginning; and, though the charter was not yet obtained, feeling assured by this official encouragement, and the public interest in the project, that it would be granted by the colony, Mr. Field remained in St. John's but three days, when he took the Merlin back to Halifax on his way to New-York, there to purchase and send down a steamer for the service of the Company, leaving his associates to secure the charter and to carry out the arrangements with the former company. To settle all these details was necessarily a work of time. First, the charter of the old "Electric Telegraph Company" had to be repealed, to clear the way for a new charter to the Company, which was to bear the more comprehensive title of "New-York, Newfoundland, *and London.*" This charter—which had been drawn with the greatest care by the counsel of the Company, while on the voyage to Newfoundland—bore on its very front the declaration that the plans of the new Company were much broader than those of the old. In the former charter, the design was thus set forth:

"The telegraph line of this company is designed to be strictly an 'Inter-Continental Telegraph.' Its termini will be New-York, in the United States, and London, in the kingdom of Great Britain; these points are to be connected by a line of electric telegraph from New-York to St. John's, Newfoundland, partly on poles, partly laid in the ground, and partly through the water, *and a line of the swiftest steamships ever built from that point to Ireland.* The trips of these steamships, it is expected, will not exceed five days, and as very little time will be occupied in transmitting messages between St. John's and New-York, the communication between the latter city and London or Liverpool, will be effected *in six days,* or less. The company will have likewise stationed at St. John's a steam yacht, for the purpose of intercepting the European and American steamships, so that no opportunity may be lost in forwarding intelligence in advance of the ordinary channels of communication."

But the charter of the " New-York, Newfoundland, and London Telegraph Company," which was now to be obtained, began by declaring, in its very first sentence: " Whereas, it is deemed advisable, to establish a line of telegraphic communication between America and Europe, by way of Newfoundland." Not a word is said of fast ships, of communications in less than six days, but every thing points to a line across the ocean. Thus one section gives authority to establish a submarine telegraph across the ocean, from Newfoundland to Ireland; another section prohibits any other company or person from touching the coast of Newfoundland or its dependencies [which includes Labrador] with a telegraphic cable or wire, from any point whatever, for fifty years; and a third section grants the Company fifty square miles of land upon the completion of the submarine line across the Atlantic.

In other respects the charter was equally liberal. It incorporated the associates for fifty years, established perfect equality in respect to corporators and officers, between citizens of the United States and British subjects, and allowed the meetings of the stockholders and directors to be held in New-York, in Newfoundland, or in London.

To obtain such concessions was a work of some difficulty and delay. The Legislature of the province were naturally anxious to scan carefully conditions that were to bind them and their children for half a century. I have now before me the papers of St. John's of that day, containing the discussions in the Legislature; and while all testify to the deep public interest in the stupendous project, they show a due care for the interests of their own colony, which they were bound to protect. At length all difficulties were removed, and the charter was passed unanimously by the Assembly, and confirmed by the Council.

This happy result was duly celebrated, in the manner which all Englishmen approve, by a grand banquet given by the commissioners of the new Company, to the members of the Assembly and other dignitaries of the colony. The report of that dinner is now before me in the St. John's papers; and it is gratifying to perceive how heartily the enterprise was welcomed by all classes; and how fond were the anticipations of the increased intercourse it would bring, and the manifold benefits it would confer on their long-neglected island.

No sooner were the papers signed, than the wheels, so long blocked, were unloosed, and the machinery began to move. Mr. White at once drew on New-York for fifty thousand dollars, and paid off all the debts of the old company. A St. John's newspaper of April 8th, 1854, amid a great deal on the subject, contains this paragraph, which is very significant of the dead state of the old company, and of the life of the new:

"The office of the new Electric Telegraph Company has been surrounded the last two or three days by the men who had been engaged the last year on the line, and who are being paid all debts, dues, and demands against the old association. We look upon the readiness with which these claims are liquidated as a substantial indication on the part of the new Company that they will complete to the letter all that they have declared to accomplish in this important undertaking."

In the early part of May, the two gentlemen who had remained behind in Newfoundland rejoined their associates in New-York, and there the charter was formally accepted and the Company organized. As all the associates had not arrived till Saturday evening, the sixth of May, and as one of them was to leave town on Monday morning, it was agreed that they should meet for organization at six o'clock of that day. At that hour they came to the house of Mr. Field's brother, and as the first rays of the morning sun streamed into the windows, the formal organization took place. The charter was accepted. the stock subscribed, and the officers chosen. Mr. Cooper, Mr. Taylor, Mr. Field, Mr. Roberts, and Mr. White were the first directors. Mr. Cooper was chosen President, Mr. White, Vice-President, and Mr. Taylor, Treasurer.

This is a short story, and soon told. It seemed a light affair, for half a dozen men thus to meet in the early morning and toss off such a business before breakfast. But what a work was that to which they thus put their hands! A capital of a million and a half of dollars was subscribed in those few minutes, and a company put in operation that was to carry a line of telegraph to St. John's, more than a thousand miles from New-York, and then to span the wild sea. Well was it that they who undertook the work did not then fully realize its magnitude, or they might have shrunk from the attempt. Well was it for them that the veil was not lifted, which shut from their eyes the long delay, the immense toil, and the heavy burdens of many wearisome years. Such a prospect might have chilled the most sanguine spirit. But a kind Providence gives men strength for their day, imposes burdens as they are able to bear them, and thus leads them on to greater achievements than they knew.

CHAPTER IV.

The Land Line begun in Newfoundland. Immense Undertaking. Four Hundred Miles of Road to be built. Two Years of Labor. First Attempt to lay a Cable across the Gulf of St. Lawrence, in 1855. Failure. Second Attempt, in 1856, which is Successful.

There is nothing in the world easier than to build a line of railroad, or of telegraph, *on paper*. You have only to take the map, and mark the points to be connected, and then with a single sweep of the pencil to draw the line along which the iron track is to run. In this airy flight of the imagination, distances are nothing. A thousand leagues vanish at a stroke. All obstacles disappear. The valleys are exalted, and the hills are made low, noble bridges span the mountain streams, and the chasms are leaped in safety by the fire-drawn cars.

Very different is it to construct a line of railroad or of telegraph *in reality;* to come with an army of laborers, with axes on their shoulders to cut down the forests, and with spades in their hands to cast up the highway. Then poetry sinks to prose, and instead of flying over the space on wings, one must traverse it on foot, slowly and with painful steps. Then nature asserts her power; and, as if resentful of the disdain with which man in his pride affected to leap over her, she piles up new barriers in his way. The mountains with their rugged sides cannot be moved out of their place. The granite rocks must be cleft in twain, to open a passage for this boasting hero, before he can begin his triumphal march. The woods seem to thicken into an impassable jungle; and the morass sinks deeper, threatening to swallow up the horse and his rider; until the rash projector is startled at his own audacity. Then it becomes a contest of forces between man and nature, in which, if he would conquer, he must fight his way. The barriers of nature cannot be lightly pushed aside, but must yield at last only to time and toil, and "man's unconquerable will."

Seldom have all these obstacles been combined in a more formidable manner to obstruct any public work, than against the attempt to build a telegraph line across the island of Newfoundland. The distance, by the route to be traversed, was over four hundred miles, and the country was a wilderness, an utter desolation. Yet through such a country, over mountain and moor, through tangled brake and rocky gorge, over rivers and through morasses, they were to build a road—not merely a line of telegraph stuck on poles, but "a good and traversable bridle-road, eight feet wide, with bridges of the same width," from end to end of the island.

But nothing daunted, the new Company undertook the great work with spirit and resolution. Gisborne had made a beginning, and got some thirty or forty miles out of St. John's. This was the easiest part of the whole route, being in the most inhabited region of the island. But here he broke down, just where it was necessary to leave civilization behind, and to plunge into the wilderness.

Intending to resume the work on a much larger scale, Mr. White, the Vice-President, was sent down to St. John's to be the General Agent of the Company; while Mr. Matthew D. Field, as a practical engineer, was to have charge of the construction of the line. The latter soon organized a force of six hundred men, which he pushed forward in detachments to the scene of operations.

And now began to appear still more the difficulties of the way. To provide subsistence at all for man and beast, it was necessary to keep near the coast, for all supplies had to be sent round by sea. Yet in following the coast line, they had to wind around bays, or to climb over headlands. If they struck into the interior, they had to cut their way through the dense and tangled wood. There was not a path to guide them, not even an Indian trail. When lost in the forest, they had to follow the compass, as much as the mariner at sea.

To keep such a force in the field, that, like an army, produced nothing, but consumed fearfully, required constant attention to the commissary department. The little steamer Victoria, which belonged to the Company, was kept constantly plying along the coast, carrying barrels of pork and potatoes, kegs of powder, pickaxes and spades and shovels, and all the implements of labor. These were taken up to the heads of the bays, and thence carried, chiefly on men's backs, over the hills to the line of the road.

In many respects, it had the features of a military expedition. It moved forward in a great camp. The men were sheltered in tents, when sheltered at all, or in small huts which they built along the road. But more often they slept on the ground. It was a wild and picturesque sight to come upon their camp in the woods, to see their fires blazing at night while hundreds of stalwart sleepers lay stretched on the ground. Sometimes, when encamped on the hills, they could be seen afar off at sea. It made a pretty picture then. But the hardy men thought little of the figure they were making, when they were exposed to the fury of the elements. Often the rain fell in torrents, and the men, crouching under their slight shelter, listened sadly to the sighing of the wind among the trees, answered by the desolate moaning of the sea.

Yet in spite of all obstacles, the work went on. All through the long days of summer, and through the months of autumn, every cove and creek along that southern coast heard the plashing of their oars, and the steady stroke of their axes resounded through the forest.

But as the season advanced, all these difficulties increased. For nearly half the year, the island is buried in snow. Blinding drifts sweep over the moors, and choke up the paths of the forest. How at such times the expedition lay floundering in the woods, or attempting still to force its way onward; what hardships and sufferings the men endured—all this is a chapter in the History of the Telegraph which has not been written, and which can never be fully told. But

>Gentlemen of England,
>Who dwell at home at ease,

and who are justly proud of the extent of their dominions, and the life and power which pervade the whole, may here find another example of the way in which grèat works are borne forward in distant parts of their empire.

But to carry out such an enterprise, requires "head-work" as well as "hand-work." Engineering in the field must be supported by financiering at home. It was here the former enterprise broke down, and now it needed constant watching to keep the wheels in steady motion. The directors in New-York found a daily demand on their attention. The minds which had grasped the large design, must now descend to an infinity of detail. They had to keep an army of men at work, at a point a thousand miles away, far beyond their immediate oversight. Drafts for money came thick and fast. To provide for all these required constant attention. How faithfully they gave to this enterprise, not only their money, but their time and thought, few will know; but we who have seen can testify. In the autumn of that year, 1854, the writer removed to the city of New-York, and was almost daily at the house of Mr. Field. Yet for months it was hardly possible to go there of an evening without finding the library occupied by the Company. Indeed, so uniformly was this the case, that "the Telegraph" began to be regarded by the family as an unwelcome intrusion, since it put an interdict on the former social evenings and quiet domestic enjoyment. The circumstance shows the ceaseless care on the part of the directors which the enterprise involved. As a witness of their incessant labors, it is due to them to bear this testimony to their patience and their fidelity.

When they began the work, they hoped to carry

the line across Newfoundland in one year, completing it in the summer of 1855. In anticipation of this, Mr. Field was sent by the Company to England at the close of 1854, to order a cable to span the Gulf of St. Lawrence, to connect Cape Ray with the island of Cape Breton. This was his first voyage across the ocean on the business of the Telegraph—to be followed by more than thirty others. In London he met for the first time Mr. John W. Brett, with whom he was to be afterward connected in the larger enterprise of the Atlantic Telegraph. Mr. Brett was the father of submarine telegraphy in Europe, though in carrying out his first projects he was largely indebted to Mr. Crampton, a well-known engineer of London, who aided him both with advice and capital. With this invaluable assistance, he had stretched two lines across the British channel. From his success in passing these waters, he believed a line might yet be stretched from continent to continent. The scientific men of England were not then educated up to that point. The bare suggestion was received with a smile of incredulity. But Mr. Brett "had faith," even at that early day, and entered heartily into the schemes of Mr. Field. To show his interest, he afterward took a few shares in the Newfoundland line—the only Englishman who had any part in this preliminary work.

The summer came, and the work in Newfoundland, though not complete, was advancing; and the cable

HISTORY OF THE ATLANTIC TELEGRAPH. 57

in England was finished and shipped on board the bark Sarah L. Bryant to cross the sea. Anticipating its arrival, the Company chartered a steamer to go down to Newfoundland to assist in its submersion across the Gulf of St. Lawrence. As yet they had had no experience in the business of laying a submarine telegraph, and did not doubt that the work could be accomplished with the greatest ease. It was therefore to be an excursion of pleasure as well as of business, and accordingly they invited a large party to go down to witness the unaccustomed spectacle.

As we chanced to be among the guests, we have the best reason to remember it. Seldom has a more pleasant party been gathered for any expedition. Representing the Company were Mr. Field, Mr. Peter Cooper, Mr. Robert W. Lowber, and Professor Morse, while among the invited guests were gentlemen of all professions—clergymen, doctors and lawyers, artists and editors. Rev. Drs. Field and Gardiner Spring, with their white hairs, were among us, and the Rev. J. M. Sherwood; Dr. Lewis A. Sayre, Bayard Taylor, the well-known traveller, Mr. Fitz-James O'Brien, and Mr. John Mullaly—the three latter gentlemen representing leading papers of New-York.* Besides these,

---

* The letters of Mr. Taylor, which first appeared in the *Tribune*, have been since collected in one of his volumes of travel. Mr. O'Brien, a very brilliant writer, who afterward fell in our civil war, fighting

3*

the party included a large number of ladies, who gave life and animation to the company.

Well do we recall the morning of our departure. It was eleven years ago—the seventh of August, 1855. Never did a voyage begin with fairer omens. It was a bright summer day. The sky was clear, and the water smooth. We were on the deck of the good ship James Adger, long known as one of the fine steamers belonging to the Charleston line. She was a swift ship, and cut the water like an arrow. Thus we sped down the bay, and turning into the ocean, skimmed along the shores of Long Island. The sea was tranquil as a lake. The whole party were on deck, scattered in groups here and there, watching the sails and the shore. A rude telegraph instrument furnished entertainment and instruction, especially as we had Pro-

---

bravely for his adopted country, furnished some spirited letters to the *Times*. But Mr. Mullaly, who appeared for the *Herald*, was the most persevering attendant on the Telegraph, and the most indefatigable correspondent. He accompanied not only this expedition, but several others. He was on board the Niagara in 1857, and again in both the expeditions of 1858; and on the final success of the cable, prepared a volume, which was published by the Appletons, giving the history of the enterprise. This contains the fullest account of all those expeditions which has been given to the public. I have had frequent occasion to refer to his book, and can bear witness to the interest of the narrative. It is written with spirit, and doubtless would have had a longer life, if the cable itself had not come to an untimely end.

fessor Morse to explain this marvellous invention, which some of us felt that we then for the first time fully understood. All day long we lingered on the deck, and here kept watching still as the sun went down in the waves, and the stars began to twinkle on the deep. It was a day not to be forgotten.

At Halifax, several of us left the ship, and came across Nova Scotia, passing through that lovely region of Acadia which Longfellow has invested with such tender interest in his poem of Evangeline. Thence we crossed the Bay of Fundy to St. John's in New-Brunswick, and returned by way of Portland.

The James Adger went on to Newfoundland, steering first for Port au Basque, near Cape Ray, where they hoped to meet the bark which was to come from England with the cable on board. To their disappointment, it had not arrived. Mr. Canning, the engineer who was to lay the cable, had come out by steamer, and was on hand, but the bark was not to be seen. Having to wait several days, and wishing to make the most of their time, they sailed for St. John's, where they were received by the Provincial Government and the people with unbounded hospitality, after which they returned to Port au Basque, and were now rejoiced to discover the little bark hidden behind the rocks. It was decided to land the cable in Cape Ray Cove. After a day or two's delay in getting the end to the shore, they started to cross the Gulf of St. Lawrence, the Adger towing the bark. The sea was calm, and though they were obliged to move slowly, yet all promised well, till they were about half-way across, when a gale arose, which pitched the bark so violently, that with its unwieldy bulk it was in great danger of sinking. After holding on for hours in the vain hope that it would abate, the captain cut the cable to save the bark; and thus, after they had paid out forty miles, it was hopelessly lost, and the Adger returned to New-York.

This loss was owing partly to the severity of the gale, and partly to the fact that the bark which had the cable on board was wholly unfitted for the purpose. It was a sailing-vessel, and had to be towed by another ship. In this way it was impossible to regulate its motion. It was too fast or too slow. It was liable to be swayed by the sea, now giving a lurch ahead, and now dragging behind. Experience showed that a cable should always be laid from a steam-vessel which could regulate its own motion, running out freely when all went smoothly, and checking its speed instantly when it was necessary to ease up the strain, or to pay out more slack to fill up the hollows of the sea.

This first loss of a submarine cable was a severe disappointment to the Company. It postponed the enterprise for a whole year. To make a new cable would require several months. The season was now so far advanced that it could not be laid before another summer. Was it strange if some of the little band began to ask if they had not lost enough, and to reason that it was better to stop where they were, than to go on still farther, casting their treasures into the sea?

But there was in that little company a spirit of hope and determination that could not be subdued; that ever cried: "Once more unto the breach, good friends!" After some deliberation, it was resolved to renew the attempt. Mr. Field again sailed for England to order another cable, which was duly made and sent out the following summer. This time, warned by experience, the Company invited no party and made no display. The cable was placed on board a steamer fitted for the purpose. It was laid without accident, and remained in perfect working order for nine years.

Meanwhile the work on land had been pushed forward without ceasing. After incredible labor, the Company had built a road and a telegraph from one end of Newfoundland to the other, four hundred miles; and, as if that were not enough, had built also another line, one hundred and forty miles in length, in the island of Cape Breton. The first part of their work was now done. The telegraph had been carried beyond the United States through the

British Provinces to St. John's in Newfoundland, a distance from New-York of over one thousand miles.

The cost of the line, thus far, had been about a million of dollars, and of this the whole burden, with but trifling exceptions, had fallen upon the original projectors—Mr. Field having put in over two hundred thousand dollars in money — and Mr. Cooper, Mr. Taylor, and Mr. Roberts each a little less. No other contributors beyond the six original subscribers had come, except Professor Morse, Mr. Robert W. Lowber, Mr. Wilson G. Hunt, and Mr. John W. Brett. The list of directors and officers remained as it was at first, except that this year, 1856, Mr. White died, and his place as director was filled by Mr. Hunt, and that Mr. Field was chosen Vice-President, and Mr. Lowber Secretary. In all the operations of the Company thus far, the various negotiations, the plan of the work, the oversight of its execution, and the correspondence with the officers and others, mainly devolved upon Mr. Field.

And so at length, after two long and weary years, these bold projectors had accomplished half their work. They had passed over the land, and under the Gulf of St. Lawrence, and having reached the farthest point of the American coast, they now stood upon the cliffs of Newfoundland, looking off upon the wide sea.

CHAPTER V.

THE DEEP-SEA SOUNDINGS. THE OLD METHOD OF BALL AND LINE. MASSEY'S INDICATOR. INVENTION OF LIEUTENANT BROOKE. CRUISE OF THE DOLPHIN IN 1853, AND OF THE ARCTIC IN 1856. THE BRAVE LIEUTENANT BERRYMAN. SOUNDINGS BY COMMANDER DAYMAN, OF THE BRITISH NAVY, IN THE CYCLOPS IN 1857. THE BED OF THE ATLANTIC. DEPTHS IN DIFFERENT PARTS. THE TELEGRAPHIC PLATEAU. SUBMARINE MOUNTAIN OFF THE COAST OF IRELAND.

> Hark! do you hear the sea?
> —KING LEAR.

WHEN a landsman, born far away among the mountains, comes down to the coast, and stands for the first time on the shore of the sea, it excites in him a feeling of awe and wonder, not unmingled with terror. There it lies, a level surface, with nothing that lifts up its head like a peak of his native hills. And yet it is so vast, stretching away to the horizon, and all over the sides of the round world; with its tides and currents that sweep from the equator to the pole; with its unknown depths and its ceaseless motion; that it is to him the highest emblem of majesty and of power—a not unworthy symbol of God himself.

In proportion to its mystery is the terror which hangs over it. A vague dread always surrounds the unknown. And what so unknown as the deep, unfathomable sea? For thousands of years the sails of ships, like winged birds, have skimmed over it, yet it has remained the one thing in nature beyond alike man's knowledge and his power.

> Man marks the earth with ruin,
> His control stops with the shore.

And the little that has been known of the ocean has been chiefly of its *surface*, of the winds that blow over it, and the waves that are lifted up on high. We knew somewhat of its tides and currents as observed in different parts of the earth. We saw off our coast the great Gulf Stream—that steady flow of waters so mighty and mysterious, which, issuing out of the tropical regions, poured its warm current, sixty miles broad, right through the cold waters of the North-Atlantic; and sweeping round, sent the airs of a softer climate over all the countries of Western Europe. Old voyagers told us of the trade-winds that blew across the Pacific, and of terrible monsoons in China and Indian seas. But all that did not reveal what was going on a hundred fathoms below the surface. These old sailors had marvellous tales of Indian pearl-divers, who, holding their breath, plunged to the depth of a few hundred feet; but they came up half-dead, with but little to tell except of the frightful monsters of the deep. The diving-bell was let down over sunken wrecks, but the divers came up only with tales of riches and ruin, of gold and gems and dead men's bones that lie mingled together on the deep sea floor. Was the bottom of the sea all like this? Was it a vast realm of death, the sepulchre of the world? No man could tell us. Poets might sing of the caves of ocean, but no eye of science had yet penetrated those awful depths, which the storms never reach.

It is indeed marvellous how little was known, up to a very recent date, of the true character of the ocean. Navigators had often tried to find out how deep it was. When lying becalmed on a tranquil sea, they had amused themselves by letting down a long line, weighted with a cannon-ball, to see if they could touch bottom. But the results were very uncertain. Sometimes the line ran out for miles and miles, but whether it was all the while descending, or was swayed hither and thither by mighty under-currents, could not be known.

But this true character of the ocean it was necessary to determine, before it could be possible to pass the gulf of the Atlantic. What was there on the bottom of the sea? for it was there the cable was to find its resting-place. Was that ocean-bed a wide level plain, or had it been heaved up by volcanic forces into a hundred mountain-peaks, with many a gorge and precipice between? Such *was* the character of a part of the basin of the ocean. Here and there, all over the globe, are islands, like the Peak of Teneriffe, thrown up in some fierce bursting of the crust of our planet, that shoot up in tremendous cliffs from the sea. Who shall say that the same cliffs do not shoot *down* below the waves a thousand fathoms deep? And might there not be such islands, which did not show their heads above the surface, lying in the track between Europe and America; or perchance a succession of mountain ranges, over which the cable would have to be stretched, and where hanging from the heights it would swing with the tide, till at last it snapped and fell into the abyss below? Such at least were possible dangers to be encountered; and it was not safe to advance a step till first the basin of the North-Atlantic was explored.

The progress of invention, so rapid on land, at length found a way of penetrating the sea, and even of turning up its bottom to the gaze of men. To measure the depth with something like mathematical accuracy, an instrument was introduced known among nautical men as Massey's Indicator, the method of which is very clearly explained in an article which appeared in one of the New-York papers, (the

Times,) on the deep-sea soundings made for the Atlantic Telegraph:

"The old system is with a small line, marked at distances of one hundred fathoms, and with a weight of thirty or fifty pounds, the depth being told by the length of line run out. This is, of course, the most natural apparatus that suggests itself, and has been in use from the earliest ages. Experience has given directions for its use, avoiding some of the grosser causes of error from driftage and other causes. Yet its success in immense ocean depths is problematical, and a problem decided in the negative by many of the first scientific authorities at home and abroad. In the mechanical improvements of the last half-century substitutes for this simple but rather uncertain method began to be devised. It was proposed to ascertain the depth by the amount of pressure, or by explosions under water, with other equally impracticable plans. At last was noticed the perfect regularity of the movements of a spirally-shaped wheel, on being drawn through the water. Experiments proved that this regularity, when unaffected by other causes, could be relied on with perfect accuracy, and that an arrangement of cog-wheels would register its revolutions with mathematical precision. Very soon it came in use as a ship's log. So perfect was their precision, that they were even introduced in scientific surveys.

Base lines, where the nicest accuracy is required, were run with them, and we have the highest authority of the Royal Navy for believing that they never failed. At this point it was proposed to apply them in a perpendicular as well as in a horizontal motion through the water, Massey's apparatus promising to solve those problems of submarine geography left unsolved by the old method of obtaining depth with a simple line and sinker, and this more especially as some causes of error, considerable on the surface, disappear in the still water below."

To make our knowledge of the sea complete, one thing more was wanting—a method not only of reaching the bottom, but of laying hold of it, and bringing it up to the light of day. This was now to be supplied.

It is to the inventive genius of a lieutenant of the United States Navy, Mr. J. M. Brooke, that the world owes the means of finding out what is at the bottom of the sea. This is by a very simple contrivance, by which the heavy weight, used to sink the measuring line, *is detached as soon as it strikes bottom*, leaving the line free so that it can be drawn up lightly and quickly to the surface without danger of breaking. Below the weight, and driven by it into the ooze, is a rod, in which is an open valve, that now closes with a spring, by which it catches a cupful of the soil, which is thus brought up to the surface, to be placed

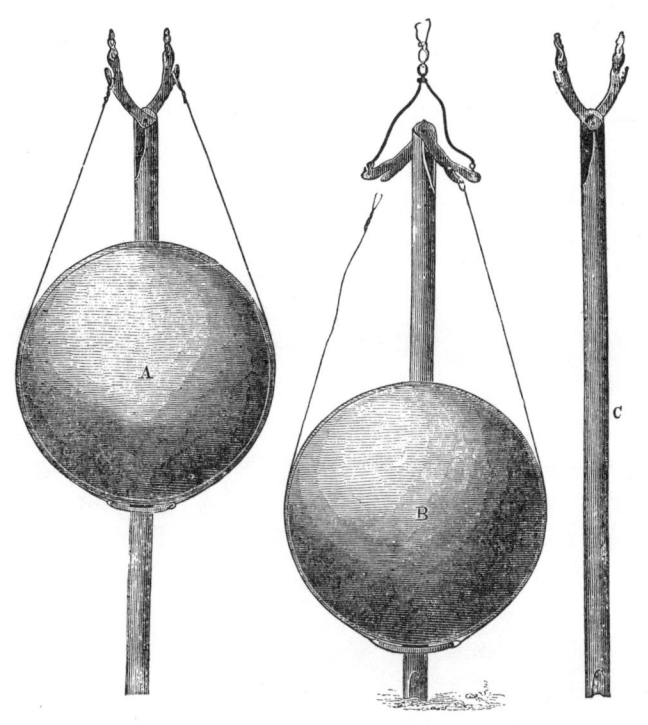

BROOKE'S DEEP SEA SOUNDING APPARATUS.

A shows the instrument ready for sounding. It is very simple, consisting only of a cannon-ball, pierced with an iron rod, and held in its place by slings. As the ball goes down swiftly, it drives the rod into the bottom like the point of a spear, when an opening at the end catches the ooze in its iron lips. The same instant (see B,) the slings loosen, the ball drops off, and the naked rod, C, with its "bite" is drawn up to the surface.

under the microscope, and be subjected to the sharp eye of science. With this simple instrument the skilful seaman explores the bottom of the ocean by literally feeling over it. With a long line he dives to the very lowest depths, while the clasp at the end of it, is like the tip of the elephant's trunk, serving as a delicate finger with which he picks up sand and shells that lie strewn on the floor of the deep. What important conclusions are derived from this inspection of the bottom of the sea, is well stated by Lieutenant Maury in the letter already quoted.

We can but regard it as a Providential event, preparing the way for the great achievement which was now to be undertaken, that a partial survey of the Atlantic had been made the very year before this enterprise was begun, in 1853. Lieutenant Berryman was the first who applied this new method of taking deep-sea soundings to that part of the Atlantic lying between Newfoundland and Ireland, with results most surprising and satisfactory. But to remove all doubt it seemed desirable to have a fresh survey. To obtain this, Mr. Field went to Washington and applied to the Government in behalf of the Company for a second expedition.

The request was granted, and the Arctic, under command of the same gallant Lieutenant Berryman, was assigned to this service. He sailed from New-York on the eighteenth of July, 1856, and the very next day Mr. Field left on the Baltic for England, to organize the Atlantic Telegraph Company. The Arctic proceeded to St. John's, and thence with a clear eye and a steady hand, this true sailor went "sounding on his dim and perilous way" across the deep. In about three weeks he made the coast of Ireland, having carried his survey along the great circle arc, which the telegraph was to follow as the nearest path from the old world to the new. The result fully confirmed his belief of the existence of a great plateau underneath the ocean, extending all the way from one hemisphere to the other.

I cannot take leave of the name of this gallant officer, who rendered such services to science and to his country, without a word of tribute to his memory. Lieutenant Berryman is in his grave. He died in the navy of his country, and from his ardent devotion to her service. When, five years ago, the great civil war which has just ended, broke out, he was placed in a position most painful to a man of large heart, who loved at once his country and the state in which he was born. He was a Southerner, a native of Winchester, Va., and was assigned to service in the South. At the first attack on Southern forts and arsenals, he was in command of the Wyandotte, in the harbor of Pensacola, in Florida. His officers, who were nearly all Southerners, were in secret sympathy with the rebellion. Thus all the influences around him, both on ship and on shore, were such as might have seduced a weaker man from his loyalty. But, to his honor, he never hesitated for a moment. He stood firm and loyal to his flag. Not knowing whom to trust, he had to keep watch day and night against surprise and treachery. It was the testimony of Lieutenant Slemmer, then in command of Fort Pickens, that but for the ceaseless exertions of Lieutenant Berryman not only the ship but the fort would have been lost. But this victory for his country he paid for with his life. His constant exertions brought on a brain fever, of which he died. His wife, also a native of Winchester, when the war came near her early home, removed to Baltimore, saying that "she would not live under any other flag than that under which her husband had lived and died."

It was to the honor of the American navy, to have led the way in these deep-sea soundings. But after this second voyage of exploration, Mr. Field applied to the British Admiralty, "to make what further soundings might be necessary between Ireland and Newfoundland, and to verify those made by Lieutenant Berryman." It was in response to this application that the Government sent out the following year a vessel to make still another survey of the same ocean-path. This was the steamer Cyclops, which was placed under Lieutenant Commander Joseph Dayman, of the British navy, an officer who had been with Captain Sir James Ross when he made his deep-sea soundings in the South-Atlantic in 1840, where he attained a depth of twenty-six hundred and sixty-seven fathoms; and who by his intelligence and zeal, was admirably fitted for the work. To speak now of this *third* survey, is anticipating in time. But it will serve the purpose of unity and clearness in the narrative, to include all these deep-sea soundings in one chapter. He was directed to proceed to the harbor of Valentia in Ireland, and thence to follow, as nearly as possible, along the arc of a great circle to Newfoundland. " The soundings for the first few miles from the coast should be frequent, decreasing as you draw off shore."

These orders were thoroughly executed. Every pains was taken to make the information obtained precise and exact. Whenever a sounding was to be taken, the ship was hove to, and the bow kept as nearly as possible in the same spot, so that the line might descend perpendicularly. This was repeated every few miles until they had got far out into the Atlantic, where the general equality of the depths rendered it necessary to cast the line only every twenty or thirty miles. Thus the survey was made complete, and the results obtained were of the greatest value in determining the physical geography of the sea.

The conclusions of Commander Dayman confirmed in general those of Lieutenant Berryman, though in comparing the charts prepared by the two, we observe some differences which ought to be noticed. Both agree as to the general character of the bottom of the ocean along this latitude—that it is a vast plain, like the steppes of Siberia. Yet on the chart of Dayman the floor of the sea seems *not such a dead level* as on that of Berryman. (This may be partly owing to a difference of route, as Dayman passed a little to the north of the track of Berryman.) There are more unequal depths, which in the small space of a chart, appear like hills and valleys. Yet when we consider the wide distances passed over, these inequalities seem not greater than the undulations on our Western prairies. "This space," says Dayman, "has been named by Maury the telegraphic plateau, and although by multiplying the soundings upon it, we have depths ranging from fourteen hundred and fifty to twenty-four hundred fathoms, these are comparatively small inequalities in its surface, and present no new difficulty to the project of laying the cable across the ocean. Their importance vanishes when the extent of the space over which they are distributed (thirty degrees of longitude) is considered."

According to Berryman and Dayman both, the ocean in its deepest part on this plateau, measured but a little over two thousand fathoms, or twelve thousand feet—a depth of not much over two miles. This is not great, compared with the enormous depths in other parts of the Atlantic;* yet that it is *something*, may be realized from the fact that if the Peak of Teneriffe were here " cast into the sea," it would sink out of sight, island, mountain and all, while even the lofty head of Mont Blanc would be lifted but a few hundred feet above the waves.

The only exception to this uniform depth, lies about two hundred miles off the coast of Ireland, where within a space of about a dozen miles, the depth sank from five hundred and fifty to seventeen hundred and

* " The ocean bed of the North-Atlantic is a curious study; in some parts furrowed by currents, in others presenting banks, the accumulations perhaps of the débris of these ocean rivers during countless ages. To the west, the Gulf Stream pours along in a bed from one mile to a mile and a half in depth. To the east of this, and south of the Great Banks, is a basin, eight or ten degrees square, where the bottom attains a greater depression than perhaps the highest peaks of the Andes or Himalaya—six miles of line have failed to reach the bottom. Taking a profile of the Atlantic basin in our own latitude, we find a far greater depression than any mountain elevation on our own continent. Four or five Alleghanies would have to be piled on each other, and on them added Fremont's Peak, before their point would show itself above the surface. Between the Azores and the mouth of the Tagus this decreases to about three miles."

fifty fathoms! "In 14° 48' west," says Dayman, "we have five hundred and fifty fathoms rock, and in 15° 6' west we have seventeen hundred and fifty fathoms ooze. This is the greatest dip in the whole ocean." "In little more than ten miles of distance a change of depth occurs, amounting to seventy-two hundred feet." This seems indeed a tremendous plunge, especially off from the hard rock into the slime of the sea.

The same sudden declivity was noticed by Berryman, and has been observed in the several attempts to lay the cable. Thus in the second expedition of 1858, as the Agamemnon was approaching the coast of Ireland, we read in the report of her voyage: "About five o'clock in the evening, the steep submarine mountain which divides the telegraphic plateau from the Irish coast, was reached; and the sudden shallowing of the water had a very marked effect on the cable, causing the strain on, and the speed of it, to lessen every minute. A great deal of slack was paid out to allow for inequalities which might exist, though undiscovered by the sounding-line."

This submarine mountain was then regarded as the chief point of danger in the whole bed of the Atlantic, and as the principal source of anxiety in laying a cable across the ocean. Yet, after all, a descent or ascent of less than a mile and a half in ten miles, is not an impassable grade. More recent soundings reduce this still farther. Captain Hoskins, R. N., has since made a second survey, and with results much more favorable. The side of the mountain, it is now said, is not very much steeper than Holborn Hill in London, or Murray Hill in New-York.* *But perhaps the best answer to fears on this point, is the fact that in 1857, 1858, and 1865, the cable passed over it without difficulty. In 1857 the Niagara was a hundred miles farther to sea, when the cable broke. In 1865 the strain was not increased more than a hundred pounds.

* The following is from a recent article in the London Times:

"The dangerous part of this course has hitherto been supposed to be the sudden dip or bank which occurs off the west coast of Ireland, and where the water was supposed to deepen in the course of a few miles from about three hundred fathoms to nearly two thousand. Such a rapid descent has naturally been regarded with alarm by telegraphic engineers, and this alarm has led to a most careful sounding survey of the whole supposed bank by Captain Dayman, acting under the instructions of the Admiralty. The result of this shows that the supposed precipitous bank, or submarine cliff, is a gradual slope of nearly sixty miles. Over this long slope the difference between its greatest height and greatest depth is only eighty-seven hundred and sixty feet; so that the average incline is, in round numbers, about one hundred and forty-five feet per mile. A good gradient on a railway is now generally considered to be one in one hundred feet, or about fifty-three in a mile; so that the incline on this supposed bank is only about three times that of an ordinary railway. In fact, as far as soundings can demonstrate any thing, there are few slopes in the bed of the Atlantic as steep as that of Holborn Hill. In no part is the bottom rocky, and with the exception of a few miles, which are shingly, only ooze, mud, or sand is to be found."

Next to the depth of the ocean, it was important to ascertain the nature of its bottom. What was it—a vast bed of rock, the iron-bound crust of the globe, hardened by internal fires, and which, bending as a vault over the still glowing centre of the earth, bore up on its mighty arches the weight of all the oceans? or was it mere sand like the sea-shore? or ooze as soft as that of a mill-pond? The pressure of a column of water two miles high would be equal to that of four hundred atmospheres. Would not this weight alone be enough to crush any substance that could reach that tremendous depth? These were questions which remained to be answered, but on which depended the possibility of laying a cable at the bottom of the Atlantic.

By the ingenious contrivance of Lieutenant Brooke, the problem was solved, for we got hold of fragments of the under-coating of the sea; and to our amazement, instead of finding the ocean bound round with thick ribs of granite, its inner lining was found to be soft as a silken vest. The soil brought up from the bottom was not even of the hardness of sand or gravel. It was mere ooze, like that of our rivers, and was as soft as the moss that clings to old, damp stones on the river's brink. At first it was thought by Lieutenant Berryman to be common clay, but being carefully preserved, and subjected to a powerful microscope, it was found to be composed of myriads of shells, too small to be discovered by the naked eye!

This fact alone was a revelation. What a story did it tell of the forms of animated existence which fill the sea. "The ocean teems with life, we know. Of the four elements of the old philosophers—fire, earth, air, and water—perhaps the sea most of all abounds with living creatures. The space occupied on the surface of our planet by the different families of animals and their remains are inversely as the size of the individual. The smaller the animal, the greater the space occupied by his remains. Take the elephant and his remains, or a microscopic animal and his, and compare them. The contrast, as to space occupied, is as striking as that of the coral reef or island with the dimensions of the whale. The graveyard that would hold the corallines is larger than the graveyard that would hold the elephants."*

These little creatures, whose remains were thus found at the bottom of the ocean, probably did not live there, for there all is dark, and shells, like flowers, need the light and warmth of the all-reviving sun. It was their sepulchre, but not their dwelling place. Probably they lived near the surface of the ocean, and after their short life, sunk to the tranquil waters below. What a work of life and death had been going on for

* Maury's Physical Geography of the Sea.

ages in the depths of the sea! Myriads upon myriads, ever since the morning of creation, had been falling like snow-flakes, till their remains literally covered the bottom of the deep.

Equally significant was the fact that these shells were *unbroken.* Not only were they there, but preserved in a perfect form. Organisms the most minute and delicate, fragile as drooping flowers, had yet sunk and slept uninjured. The same power which watches over the fall of a sparrow had kept these frail and tender things, and after their brief existence, had laid them gently on the bosom of the mighty mother for their eternal rest.

The bearing of this discovery on the problem of a submarine telegraph was obvious. For it too was to lie on the ocean-bed, beside and among these relics that had so long been drifting down upon the watery plain. And if these tiny shells slept there unharmed, surely an iron chord might rest there in safety. There were no swift currents down there; no rushing waves agitated that sunless sea. There the waters moved not; and there might rest the great nerve that was to pass from continent to continent. And so far as injury from the surrounding elements was concerned, there it might remain, whispering the thoughts of successive generations of men, till the sea should give up its dead.

CHAPTER VI.

Mr. Field goes to England to organize the Atlantic Telegraph Company. Confers with John W. Brett. Seeks Counsel of Engineers and Electricians. Result of Experiments. Applies to the Government for aid. Letter from the Treasury. Enters into an agreement with Messrs. Brett, Bright, and Whitehouse to form a Company. The Enterprise brought before the British Public. Capital raised and Company organized. Choice of a Board of Directors. Contract for the Cable.

Up to this time the Telegraph, which was destined to pass the sea, had been purely an American enterprise. It had been begun, and for two years had been carried on, wholly by American capital. Save the few shares held by Mr. Brett, which are hardly enough to be counted an exception, not a dollar had been raised on the other side of the Atlantic. Indeed we might add, not a dollar had been raised on *this*, outside of the little circle in which the scheme had its origin. No stock or bonds were put upon the market; no man was asked for a subscription. If they wanted money, they drew their checks for it. At one time, indeed, two hundred and fifty thousand dollars of bonds were issued, but they were at once taken wholly by themselves. It is but just that those who thus single-handed bore the burden, and who risked a total loss in the event of failure, should have the honor in the event of success.

But, as the time was now come when the long-meditated attempt was to be made to carry the Telegraph across the ocean, it was fitting that Great Britain, whose shores it was to touch, should join in the work. Accordingly, in the summer of 1856, after finishing all that he could do in America, Mr. Field sailed with his family for England. The very day before he embarked, he had the pleasure to see his friend, Lieutenant Berryman, off on his second voyage to make soundings across the Atlantic.

In London he sought at once Mr. John W. Brett, with whom in his two former visits to England he had already discussed the project of a Telegraph across the ocean, and found him prompt with his counsel and coöperation. As we go on with this history, it is a melancholy satisfaction to refer to one and another worker in this enterprise, who lived not to see its last and greatest triumph. Mr. Brett, like Berryman, is dead. But he did not go to his grave till after a life of usefulness and honor. He was one of the men of the new era—of the school of Stephenson and Brunel—who believed in the marvellous achievements yet to be wrought by human invention, turning to the

*4

service of man the wonders of scientific discovery. He was one of the first to see the boundless possibilities of the telegraph, and to believe that what had passed over the land might pass under the sea. He was the first to lay a cable across the British Channel, and thus to bring into instantaneous communication the two great capitals of Europe — an achievement which, though small compared with what has since been done, was then so marvellous, that the intelligence of its success was received with surprise and incredulity. Many could not and would not believe it. Even after messages were received in London from Paris, there were those who declared that it was an imposition on the public, with as much obstinacy as some still persist that a message never passed over the Atlantic Telegraph!

This friendship of Mr. Brett—both to the enterprise and to himself personally—remained to the last. In every voyage to England Mr. Field found—however others doubted or despaired—that Mr. Brett was always the same—full of hope and confidence. Only two years ago, when they met in London, he was unshaken in faith, and urgent to have the great enterprise renewed. That he was not to see. But, though he has passed away, his work remains; and whenever two hemispheres shall rejoice over a fresh triumph of the Atlantic Telegraph, they should remember his eminent services. And therefore do we pause at his name—to lay our humble tribute on the grave of this true-hearted Englishman.

To Mr. Brett, therefore, he went first to consult in regard to his great project of a telegraph across the Atlantic Ocean. This was a part of the design embraced in the original organization of the New-York, Newfoundland, and London Telegraph Company; and when Mr. Field went to England, he was empowered to receive subscriptions to that Company, so as to enlarge its capital, and thus include in one corporation the whole line from New-York to London; or to organize a new company, which should lay a cable across the Atlantic, and there join the Newfoundland line.

But before an enterprise so vast and so new could be commended to the commercial public of Great Britain, there were many details to be settled. The mechanical and scientific problems already referred to, whether a cable could be laid across the ocean; and if so, whether it could be *worked*, were to be considered anew. The opinions of Lieutenant Maury and of Professor Morse were published in England, and arrested the attention of scientific men. But John Bull is slow of belief, and asked for more evidence. The thing was too vast to be undertaken rashly. As yet there was no experience to decide the possibility of a telegraph across the ocean. *The longest line which had been laid was three hundred miles.* This caution, which is a national trait of Englishmen, will not be regarded as a fault by those who consider that in proportion as they are slow to embark in any new enterprise, are they resolute and determined in carrying it out.

To resolve these difficult problems, Mr. Field sought counsel of the highest engineering authorities of Great Britain, and of her most eminent scientific men. To their honor, all showed the deepest interest in the project, and gave it freely the benefit of their knowledge.

First, as to the possibility of laying a cable in the deep sea, Mr. Field had witnessed one attempt of the kind—that in the Gulf of St. Lawrence the year before—an attempt which had failed. His experience, therefore, was not very encouraging. If they found so much difficulty in laying a cable seventy miles long, how could they hope to lay one of two thousand miles across the stormy Atlantic?

This was a question for the Engineers. To solve the problem, required experiments almost without number. It was now that the most important services were rendered by Glass, Elliot & Co., of London, a firm which had begun within a few years the manufacture of seacables, and which was to write its name in all the waters of the world. Aided by the skill of their admirable engineer, Mr. Canning, they now manufactured cables almost without end, applying to them every possible test. At the same time, Mr. Field took counsel of Robert Stephenson and George Parker Bidder, both of whom manifested a deep interest in the success of the enterprise.

Not less cordial was Mr. Brunel, a name known in both hemispheres. From the beginning he showed the warmest interest in the undertaking, and made many suggestions in regard to the form of the cable, and the manner in which it should be laid. He was then building the Great Eastern; and one day he took Mr. Field down to Blackwall to see it, and said, pointing to the monstrous hull which was rising on the banks of the Thames: "There is the ship to lay the Atlantic cable!" Little did he think that ten years after, that ship would be employed in this service; and in this final victory over the sea, would redeem all the misfortunes of her earlier career.

Among the difficulties, partly mechanical and partly scientific, to be encountered, was that of finding a perfect insulator. Without insulation, telegraphic communication by electricity is impossible. On land, where wires are carried on the tops of poles, the air itself is a sufficient insulator. A few glass rings at the points where the wire passes through the iron staples by which it is supported, and the insulation is complete. But in the sea the electricity would be instantly dissipated, unless some material could be found which should insulate a conductor sunk in water, as completely as if it were raised in air. But what could thus inclose the lightning, and keep it fast while flying from one continent to the other?

Here again it seemed as if Divine wisdom had anticipated the coming of this great enterprise, and provided in the realm of nature every material needed for its success. It was one of the remarkable, or rather should we say, Providential discoveries, which prepared the way for this final achievement, that only a few years before there should be found, in the forests of the Malayan archipelago, a substance till then unknown to the world, but which answered completely this new demand of science. This was Gutta-Percha, a substance impenetrable by water, and which is at the same time a bad conductor of electricity; so that it forms at once a perfect protection and insulation to a telegraph passing through the sea. In the experiments that were made to test the value of this material in the grander use to which it was to be applied, no man rendered heartier service, or showed a more enlightened zeal, than Mr. Samuel Statham, of the London Gutta-Percha Works—a name to be always gratefully remembered in the early history of the Atlantic Telegraph.

The mechanical difficulties removed, and the insulation provided, there remained yet the great scientific problem: Could a message be sent two thousand miles under the Atlantic? The ingenuity of man might devise some method of laying a cable across the sea, but of what use were it, if the lightning should shrink from the dark abyss?

It was in prosecuting inquiries to resolve this problem, that Mr. Field became acquainted with two gentlemen who were to be soon after associated with him in the organization of the Atlantic Telegraph Company. These were Mr. Charles T. Bright, afterward knighted for his part in laying the Atlantic cable in 1858, and Dr. Edward O. W. Whitehouse, both favorably known in England, the former as an engineer, and the latter for his many experiments in electromagnetism, as applied to the business of telegraphing. He had invented an instrument by which to ascertain and register the velocity of electric currents through submarine cables. Both these gentlemen were full of the ardor of science, and entered on this new project with the zeal which the prospect of so great a triumph might inspire. With them was now to be associated our distinguished countryman, Professor Morse. It was a most fortunate circumstance for the new enterprise that he was at this time in London, and gave his invaluable aid to the experiments which were made to determine the possibility of telegraphic communication at great distances under the sea. The result of his experiments he communicates in a letter to Mr. Field:

"LONDON, FIVE O'CLOCK A.M.,
"October 3, 1856.

"MY DEAR SIR: As the electrician of the New-York, Newfoundland, and London Telegraph Company, it is with the highest gratification that I have to apprise you of the result of our experiments of this morning upon a single continuous conductor of more than two thousand miles in extent, a distance you will perceive sufficient to cross the Atlantic Ocean, from Newfoundland to Ireland.

"The admirable arrangements made at the Magnetic Telegraph Office in Old Broad street, for connecting ten subterranean gutta-percha insulated conductors, of over two hundred miles each, so as to give one continuous length of more than two thousand miles during the hours of the night, when the telegraph is not commercially employed, furnished us the means of conclusively settling, by actual experiment, the question of the practicability as well as the practicality* of telegraphing through our proposed Atlantic cable.

---

* Professor Morse is fond of the distinction between the words practical and practicable. A thing may be practicable, that is, possible of accomplishment, when it is not a practical enterprise, that is, one which can be worked to advantage. He here argues that the Atlantic Telegraph is both practicable or possible, and at the same time a wise, practical undertaking.

"This result had been thrown into some doubt by the discovery, more than two years since, of certain phenomena upon subterranean and submarine conductors, and had attracted the attention of electricians, particularly of that most eminent philosopher, Professor Faraday, and that clear-sighted investigator of electrical phenomena, Dr. Whitehouse; and one of these phenomena, to wit, the perceptible retardation of the electric current, threatened to perplex our operations, and required careful investigation before we could pronounce with certainty the commercial practicability of the Ocean Telegraph.

"I am most happy to inform you that, as a crowning result of a long series of experimental investigation and inductive reasoning upon this subject, the experiments under the direction of Dr. Whitehouse and Mr. Bright, which I witnessed this morning—in which the induction coils and receiving magnets, as modified by these gentlemen, were made to actuate one of my recording instruments — have most satisfactorily resolved all doubts of the practicability as well as practicality of operating the telegraph from Newfoundland to Ireland.

"Although we telegraphed signals at the rate of two hundred and ten, two hundred and forty-one, and, according to the count at one time, even of two hundred and seventy per minute upon my telegraphic register, (which speed, you will perceive, is at a rate commercially advantageous,) these results were accomplished notwithstanding many disadvantages in our arrangements of a temporary and local character—disadvantages which will not occur in the use of our submarine cable.

"Having passed the whole night with my active and agreeable collaborators, Dr. Whitehouse and Mr. Bright, without sleep, you will excuse the hurried and brief character of this note, which I could not refrain from sending you, since our experiments this morning settle the scientific and commercial points of our enterprise satisfactorily.

"With respect and esteem, your obedient servant,

"SAMUEL F. B. MORSE.

"To CYRUS W. FIELD, Esq., *Vice-President of the New-York, Newfoundland, and London Telegraph Company,* 37 *Jermyn street, St. James's street.*"

The following, written a week later, confirms the impressions of the former:

"LONDON, October 10, 1856.

"MY DEAR SIR: After having given the deepest consideration to the subject of our successful experiments the other night, when we signalled clearly and rapidly through an unbroken circuit of subterranean conducting wire, over two thousand miles in length, I sit down to give you the result of my reflections and calculations.

"There can be no question but that, with a cable containing a single conducting wire, of a size not exceeding that through which we worked, and with equal insulation, it would be easy to telegraph from Ireland to Newfoundland at a speed of at least from eight to ten words per minute; nay, more: the varying rates of speed at which we worked, depending as they did upon differences in the arrangement of the apparatus employed, do of themselves prove that even a higher rate than this is attainable. Take it, however, at ten words in the minute, and allowing ten words for name and address, we can safely calculate upon the transmission of a twenty-word message in three minutes;

"Twenty such messages in the hour;

"Four hundred and eighty in the twenty-four hours, or fourteen thousand four hundred words per day.

"Such are the capabilities of a single wire cable fairly and moderately computed.

"It is, however, evident to me, that by improvements in the arrangement of the signals themselves, aided by the adoption of a code or system constructed upon the principles of the best nautical code, as suggested by Dr. Whitehouse, we may at least double the speed in the transmission of our messages.

"As to the structure of the cable itself, the last specimen which I examined with you seemed to combine so admirably the necessary qualities of strength, flexibility, and lightness, with perfect insulation, that I can no longer have any misgivings about the ease and safety with which it will be submerged.

"In one word, the doubts are resolved, the difficulties overcome, success is within our reach, and the great feat of the century must shortly be accomplished.

"I would urge you, if the manufacture can be completed within the time, (and all things are possible now,) to press forward the good work, and not to lose the chance of laying it during the ensuing summer.

"Before the close of the present month, I hope to be again landed safely on the other side of the water, and I full well know, that on all hands the inquiries of most interest with which I shall be met, will be about the Ocean Telegraph.

"Much as I have enjoyed my European trip this year, it would enhance the gratification which I have derived from it more than I can describe to you, if on my return to America, I could be the first bearer to my friends of the welcome intelligence that the great work had been begun, by the commencement of the manufacture of the cable to connect Ireland with the line of the New-York, Newfoundland, and London

Telegraph Company, now so successfully completed to St. John's.

"Respectfully, your obedient servant,
             "SAMUEL F. B. MORSE.
"To CYRUS W. FIELD, Esq., *Vice-President*, *etc.*

These experiments and others removed the doubts of scientific men. Professor Faraday, in spite of the law of the retardation of electricity on long circuits, which it was said he had discovered, and which would render it impossible to work a line of such length as from Ireland to Newfoundland, now declared his full conviction that it was within the bounds of possibility. The passage of electricity might not be absolutely instantaneous, or have the swiftness of the solar beam in its flight from the morning sun, yet it would be rapid enough *for all practical purposes.* When Mr. Field asked him how long it would take for the electricity to pass from London to New-York, he answered: "Possibly one second!"

Being thus fortified by the highest scientific and engineering authorities, the projectors of an ocean telegraph were now ready to bring it before the British public, and to see what support could be found for the undertaking from the English Government and the English people.

Mr. Field first addressed himself to the Government. Without waiting for the Company to be fully organized, with true American eagerness and impatience, he wrote a letter to the Admiralty asking for a fresh survey of the route to be traversed, and for the aid of Government ships to lay the cable. He also addressed a letter to Lord Clarendon, stating the large design which they had conceived, and asking for it the aid which was due to what concerned the honor and interest of England. The reply was prompt and courteous, inviting him to an interview for the purpose of a fuller explanation. Accordingly, Mr. Field, with Professor Morse, called upon him at the Foreign Office, and spent an hour in conversation on the proposed undertaking. Lord Clarendon showed great interest, and made many inquiries. He was a little startled at the magnitude of the scheme, and the confident tone of the projectors, and asked pleasantly: "But, suppose you *don't* succeed? Suppose you make the attempt and fail—your cable is lost in the sea—then what will you do?" "Charge it to profit and loss, and go to work to lay another," was the quick answer of Mr. Field, which amused him as a truly American reply. In conclusion, he desired him to put his request in writing, and, without committing the Government, encouraged him to hope that Britain would do all that might justly be expected in aid of this great international work. How nobly this promise was kept, time will show.

While engaged in these negotiations, Mr. Field took his family to Paris, and there met with a great loss in the sudden death of a favorite sister, who had accompanied them abroad. Full of the sorrow of this event, and unfitted for business of any kind, he returned to London to find an invitation to go into the country and spend a few days with Mr. James Wilson, then Secretary to the Treasury, and a man of great influence in the British Government, at his residence near Bath; there to discuss quietly and at length the proposed aid of the Government to the Atlantic telegraph. Though he had but little spirit to go among strangers, he felt it his duty not to miss such an opportunity to advance the cause he had so much at heart. Accordingly he went; and the result of this visit was the following letter, received a few days later:

"TREASURY CHAMBERS, Nov. 20, 1856.

"SIR: Having laid before the Lords Commissioners of her Majesty's Treasury your letter of the 13th ultimo, addressed to the Earl of Clarendon, requesting, on behalf of the New-York, Newfoundland, and London Telegraph Company, certain privileges and protection in regard to the line of telegraph which it is proposed to establish between Newfoundland and Ireland, I am directed by their lordships to acquaint you that they are prepared to enter into a contract with the said Telegraph Company, based upon the following conditions, namely:

"1. It is understood that the capital required to lay down the line will be (£350,000) three hundred and fifty thousand pounds.

"2. Her Majesty's Government engage to furnish the aid of ships to take what soundings may still be considered needful, or to verify those already taken, and favorably to consider any request that may be made to furnish aid by their vessels in laying down the cable.

"3. The British Government, from the time of the completion of the line, and so long as it shall continue in working order, undertakes to pay at the rate of (£14,000) fourteen thousand pounds a year, being at the rate of four per cent on the assumed capital, as a fixed remuneration for the work done on behalf of the Government, in the conveyance outward and homeward of their messages. This payment to continue until the net profits of the Company are equal to a dividend of six pounds per cent, when the payment shall be reduced to (£10,000) ten thousand pounds a year, for a period of twenty-five years.

"It is, however, understood that if the Government messages in any year shall, at the usual tariff-rate charged to the public, amount to a larger sum, such additional payment shall be made as is equivalent thereto.

"4. That the British Government shall have a priority in the conveyance of their messages over all others, subject to the exception only of the Government of the United States, in the event of their entering into an arrangement with the Telegraph Company similar in principle to that of the British Government, in which case the messages of the two Governments shall have priority in the order in which they arrive at the stations.

"5. That the tariff of charges shall be fixed with the consent of the Treasury, and shall not be increased, without such consent being obtained, as long as this contract lasts.

"I am, sir, your obedient servant,

"JAMES WILSON.

"CYRUS W. FIELD, Esq., 37 Jermyn street."

With this encouragement and promise of aid, the projectors of a telegraph across the ocean now went forward to organize a company to carry out their design. Mr. Field, on arriving in England, had entered into an agreement with Mr. Brett to join their efforts for this purpose. With them were afterward united two others—Sir Charles Bright, as engineer, and Dr. Whitehouse, as electrician. These four gentlemen had entered into a formal agreement to use their exertions to form a new company, to be called The Atlantic

Telegraph Company, the object of which should be "to continue the existing line of the New-York, Newfoundland, and London Telegraph Company to Ireland, by making or causing to be made a submarine telegraph cable for the Atlantic."

As they were now ready to introduce the enterprise to the British public, Mr. Field issued a circular in the name of the Newfoundland Company, and as its Vice-President, setting forth the great importance of telegraphic communication between the two hemispheres.

The next step was to raise the capital. After the most careful estimates, it was thought that a cable could be made and laid across the Atlantic for £350,000. This was a large sum to ask from a public slow to move, and that lends a dull ear to all new schemes. But armed with facts and figures, with maps and estimates, with the opinions of engineers and scientific men, they went to work, not only in London, but in other parts of the kingdom. Mr. Field, in company with Mr. Brett, made a visit to Liverpool and Manchester, to address their Chambers of Commerce. I have now before me the papers of those cities, with reports of the meetings held and the speeches made, which show the vigor with which they pushed their enterprise. This energy was rewarded with success. The result justified their confidence. In a few weeks the whole capital was subscribed. It had been divided into three hundred and fifty shares of a thousand pounds each. Of these, a hundred and one were taken in London, eighty-six in Liverpool, thirty-seven in Glasgow, twenty-eight in Manchester, and a few in other parts of England. The grandeur of the design attracted public attention, and some subscribed solely from a noble wish to take part in such a work. Among these were Mr. Thackeray and Lady Byron. Mr. Field subscribed £100,000, and Mr. Brett £25,000. But when the books were closed, it was found that they had more money subscribed than they required, so that in the final division of shares, there were allotted to Mr. Field eighty-eight, and to Mr. Brett twelve. Mr. Field's interest was thus one fourth of the whole capital of the Company.

In taking so large a share, it was not his intention to carry this heavy load alone. It was too large a proportion for one man. But he took it for his countrymen. He thought one fourth of the stock should be held in this country, and did not doubt, from the eagerness with which three fourths had been taken up in England, that the remainder would be at once subscribed in America. Had he been able, on his return, to attend to his own interests in the matter, this expectation might have been realized; but, as we shall see, hardly did he set foot in New-York, before he was obliged to hurry off to Newfoundland on the business of the Company, and when he returned the interest had subsided, so that it required very great exertions, continued through many months, to dispose of twenty-seven shares. Thus he was by far the largest stockholder in England or America—his interest being over seven times that of Mr. Brett, who was the largest next to himself—and being more than double the amount held by all the other American shareholders put together. This was at least giving pretty substantial proof of his own faith in the undertaking.

But some may imagine that after all this burden was not so great as it seemed. In many stock companies an evil custom obtains of assigning to the projectors a certain portion of the stock as a bonus for getting up the company, which amount appears among the subscriptions to swell the capital. It is indeed subscribed, *but not paid.* So some have asked whether this large subscription of Mr. Field was not in part at least merely nominal? To this we answer, that a consideration *was* granted to Mr. Field and his associates for their services in getting up the Company, and for their exclusive rights, but this was a contingent interest in the profits of the enterprise, *to be allowed only after the cable was laid.* So that the whole amount here subscribed was a *bona-fide* subscription, and paid in solid English gold. We have now before us the receipts of the bankers of the Company for the whole amount, eighty-eight thousand pounds sterling.

The capital being thus raised, it only remained to complete the organization of the Company by the choice of a Board of Directors, and to make a contract for the cable. The Company was organized in December, 1856, by the choice of Directors chiefly from the leading bankers and merchants of London and Liverpool. The list included such honored names as Samuel Gurney, T. H. Brooking, John W. Brett, and T. A. Hankey, of London; Sir William Brown, Henry Harrison, Edward Johnston, Robert Crosbie, George Maxwell, and C. W. H. Pickering, of Liverpool; John Pender and James Dugdale, of Manchester; and Professor William Thomson, LL.D., of Glasgow, one of the most eminent men in this department of Science in Europe. Associated with these English Directors were two of our own countrymen, Mr. George Peabody and Mr. C. M. Lampson, who, residing abroad for more than a third of a century, have done much in the commercial capital of the world to support the honor of the American name. Mr. Peabody's firm subscribed £10,000, and Mr. Lampson £2000. The latter gave more time than any other Director in London, except Mr. Brooking, the second Vice-Chairman, who, however, retired from the Company after the first failure in 1858, when Mr. Lampson was chosen to fill his place. The zeal and energy of this gentleman deserve high praise. Occupied with very large business concerns of his own, he yet found time to be present at almost every meeting. But to do full justice, we should need to speak of the services of all the Directors. The whole Board was animated with the same spirit. All gave their services without compensation, and their courage bore up under repeated disasters. Never did a nobler body of men have a great enterprise committed to their trust.

It was the good fortune of the Company to have, from the beginning, in the important position of Secretary a gentleman admirably qualified for the post. This was Mr. George Saward—a name familiar to all who have followed the fortunes of the telegraph, in England or America, since he has been the organ of communication with the press and the public; and with whom none ever had occasion to transact business without recognizing his rare intelligence and courtesy.

The Company being thus in working order, proceeded to form a contract for the manufacture of a cable to be laid across the Atlantic. For many months the proper form and size of the cable had been the subject of constant experiments. The conditions were: to combine the greatest degree of strength with lightness and flexibility. It must be strong, or it would snap in the process of laying. Yet it would not do to have it too large, for it would be unmanageable. Mr. Brett had already lost a cable in the Mediterranean chiefly from its bulk. Its size and stiffness made it hard to unwind it, while its enormous weight, when once it broke loose, caused it to run out with fearful velocity, till it was soon lost in the sea. It was only the year before that this accident had occurred. It was in September, 1855, in laying the cable from Sardinia to Algeria. All was going on well, until suddenly, "about two miles, weighing sixteen tons, flew out with the greatest violence in four or five minutes, flying round even when the drums were brought to a dead stop, creating the greatest alarm for the safety of the men in the hold and for the vessel." This was partly owing to the character of the submarine surface over which they were passing. The bottom of the Mediterranean is volcanic, and is broken up into mountains and valleys. The cable, doubtless, had just passed over some Alpine height, and was now descending into some awful depth below; but chiefly it was owing to the fact that the great size and bulk of the cable made it unmanageable. This was a warning to the Atlantic Company. The point to be aimed at was to combine the flexibility of a common ship's rope with the tenacity of iron. These conditions were thought to be united in the form that was adopted.* A contract was at once made for the manu-

* FORM AND STRUCTURE OF THE CABLE.—On his return to America, Mr. Field published a letter, in answer to many inquiries addressed to him, in which he says:

"No particular connected with this great project has been the subject of so much comment through the press as the form and structure of the telegraph cable. It may be well believed that the Directors have not decided upon a matter so all-important to success, without availing themselves of the most eminent talent and experience which could be commanded. The practical history of submarine telegraphs dates from the successful submersion of the cable between Dover and Calais in 1851, and advantage has been taken of whatever instruction this history could furnish or suggest. Of the submarine cables heretofore laid down, without enumerating others, it may be interesting to mention that the one between Dover and Calais weighs six tons to the mile; that between Spezzia and Corsica, eight tons to the mile; the wire laid from Varna to Balaklava, and used during the late war, less than three hundred pounds to the mile; while the weight of the cable decided on for the Atlantic Telegraph is between nineteen hundred pounds and one ton to the mile. This cable, to use the words of Dr. Whitehouse, 'is the result of many months' thought, experiment, and trial. Hundreds of specimens have been made, comprising every variety of form, size, and structure, and most severely tested as to their powers and capabilities; and the result has been the adoption of this, which we know to possess all the properties required, and these in a far higher degree than any cable that has yet been laid. Its flexibility is such as to make it as manageable as a small line, and its strength such that it will bear, in water, over six miles of its own weight suspended vertically.' The conducting medium consists not of one single straight copper-wire, but of seven wires of copper of the best quality, twisted round each other

facture of the cable, one half being given to Messrs. Glass, Elliot & Co., of London, and the other to Messrs. R. S. Newall & Co., of Liverpool. The whole was to be completed by the first of June, ready to be submerged in the sea. The company was organized on the ninth of December, and the very next day Mr. Field sailed for America, reaching New-York on the twenty-fifth of December, after an absence of more than five months.

spirally, and capable of undergoing great tension without injury. This conductor is then enveloped in three separate coverings of gutta-percha, of the best quality, forming the core of the cable, round which tarred hemp is wrapped, and over this, the outside covering, consisting of eighteen strands of the best quality of iron-wire; each strand composed of seven distinct wires, twisted spirally, in the most approved manner, by machinery specially adapted to the purpose. The attempt to insulate more than one conducting-wire or medium would not only have increased the chances of failure of all of them, but would have necessitated the adoption of a proportionably heavier and more cumbrous cable. The tensile power of the outer or wire covering of the cable, being very much less than that of the conductor within it, the latter is consequently protected from any such strain as can possibly rupture it or endanger its insulation without an entire fracture of the cable."

CHAPTER VII.

Mr. Field returns to America. Starts immediately for Newfoundland. Returns and goes to Washington, to seek Aid from the American Government. Opposition in Congress. The Cable among the Politicians. Debate in the Senate. Support of Mr. Seward and Mr. Rusk. Bill Passed.

WHEN Mr. Field reached home from abroad, he hoped for a brief respite. He had had a pretty hard campaign during the summer and autumn in England, and needed at least a few weeks of rest; but even that was denied him. He landed in New-York on Christmas Day, and was not allowed even to spend the New Year with his family. There were interests of the Company in Newfoundland which required immediate attention, and it was important that one of the Directors should go there without delay. As usual, it devolved upon him. He left at once for Boston, where he took the steamer to Halifax, and thence to St. John's. Such a voyage may be very agreeable in summer, but in mid-winter it is not a pleasant thing to face the storms of those northern latitudes. The passage was unusually tempestuous. At St. John's he broke down, and was put under the care of a physician. But he did not stop to think of himself. The work for which he came was done; and though the physician declared it a great risk to leave his bed, he took the steamer on her return, and was again in New-York after a month's absence—a month of hardship, of exposure, and of suffering, such as he had long occasion to remember.

The 'mention of this voyage came up a year afterward at a meeting of the Atlantic Telegraph Company in London, when a resolution was offered, tendering Mr. Field a vote of thanks for "the great services he had rendered to the Company by his untiring zeal, energy, and devotion." Mr. Brooking, the Vice-Chairman, had spent a large part of his life in Newfoundland, and knew the dangers of that inhospitable coast, and in seconding the resolution he said:

"It is now about a year and a half ago since I had the pleasure of making the acquaintance of my friend Mr. Field. It was he who initiated me into this Company, and induced me to take an interest in it from its earliest stage. From that period to the present I have observed in Mr. Field the most determined perseverance, and the exercise of great talent, extraordinary assiduity and diligence, coupled with an amount of fortitude which has seldom been equalled. I have known him cross the Atlantic in the depth of winter, and, within twenty-four hours after his arrival in New-

York, having ascertained that his presence was necessary in a distant British colony, he has not hesitated at once to direct his course thitherward. That colony is one with which I am intimately acquainted, having resided in it for upward of twenty years, and am enabled to speak to the hazards and danger which attend a voyage to it in winter. Mr. Field no sooner arrived at New-York, in the latter part of December, than he got aboard a steamer for Halifax, and proceeded to St. John's, Newfoundland. In three weeks he accomplished there a very great object for this Company. He procured the passage of an Act of the Legislature which has given to our Company the right of establishing a footing on those shores,* which ere long, I hope, will result in connecting us with Ireland. That is only one of the acts which he has performed with a desire to promote the interests of this great enterprise," etc.

The very next day after his return from Newfoundland, Mr. Field was called to Washington, to seek the aid of his own Government to the Atlantic Telegraph. The English Government had proffered the most generous aid, both in ships to lay the cable, and in an annual subsidy of £14,000. It was on every account desirable that this should be met by corresponding

* The rights before conferred, it would seem, applied only to the Newfoundland Company.

liberality on the part of the American Government. Before he left England, he had sent home the letter received from the Lords Commissioners of the Treasury; and thereupon the Directors of the New-York, Newfoundland, and London Telegraph Company had inclosed a copy to the President, with a letter asking for the same aid in ships, and in an annual sum of $70,000, [equivalent to £14,000,] to be paid for the government messages, the latter to be conditioned on the success of the telegraph, and to be continued only so long as it was in full operation. They urged with reason that the English Government had acted with great liberality—not only toward the enterprise, but toward our own Government. Although both ends of the line were in the British possessions, it had claimed no exclusive privileges, but had stipulated for perfect equality between the United States and Great Britain. The agreement expressly provided "that the British Government shall have a priority in the conveyance of their messages over all others, *subject to the exception only of the Government of the United States*, in the event of their entering into an arrangement with the Telegraph Company similar in principle to that of the British Government, in which case the messages of the two governments shall have priority in the order in which they arrive at the stations."

The letter to the President called attention to this generous offer—an offer which it was manifestly to the advantage of our Government to accept—and added: "The Company will enter into a contract with the Government of the United States on the same terms and conditions as it has made with the British Government." They asked for the same recognition and aid in the United States which they had received in England. This surely was not a very extravagant request. It was natural that American citizens should think that in a work begun by Americans, and of which, if successful, their country would reap largely the honor and the advantage, they might expect the aid from their own Government which they had already received from a foreign power. It was, therefore, not without a mixture of surprise and mortification that they learned that the proposal in Congress had provoked a violent opposition, and that the bill was likely to be defeated. Such was the attitude of affairs when Mr. Field returned from Newfoundland, and which led him to hasten to Washington.

He now found that it was much easier to deal with the English than with the American Government. Whatever may be said of the respective methods of administration, it must be confessed that the forms of England furnish greater facility in the despatch of business. A contract can be made by the Lords of the

Treasury without waiting the action of Parliament. The proposal is referred to two or three intelligent officers of the Government—perhaps even to a single individual—on whose report it takes action without further delay. Thus it is probable that the action of the British Government was decided wholly by the recommendation of Mr. Wilson, formed after the visit of Mr. Field.

But in our country we do things differently. Here it would be considered a stretch of power for any administration to enter into a contract with a private company—a contract binding the Government for a period of twenty-five years, and involving an annual appropriation of money—without the action of Congress. This is a safeguard against reckless and extravagant expenditure, but, as one of the penalties we pay for our more popular form of government, in which every thing has to be referred to the people, it involves delay, and sometimes the defeat of wise and important public measures.

Besides—shall we confess it to our shame—another secret influence often appears in American legislation, which has defeated many an act demanded by the public good—the influence of the Lobby. This now began to show itself in opposition. It had been whispered in Washington that the gentlemen in New-York who were at the head of this enterprise *were very rich;* and a measure coming from such a source surely ought to be made to pay tribute before it was allowed to pass. This was a new experience. Those few weeks in Washington were worse than among the icebergs off the coast of Newfoundland. The Atlantic Cable has had many a kink since, but never did it seem to be entangled in such a hopeless twist, as when it got among the politicians.

But it would be very unjust to suppose that there were no better influences in our Halls of Congress. There were then—as there have always been in our history—some men of large wisdom and of a noble patriotic pride, who in such a measure thought only of the good of their country and of the triumph of science and of civilization.

Two years after — in August, 1858 — when the Atlantic Telegraph proved at last a reality, and the New World was full of its fame, Mr. Seward, in a speech at Auburn, thus referred to the ordeal it had to pass through in Congress:

"The two great countries of which I have spoken, [England and America,] are now ringing with the praises of Cyrus W. Field, who chiefly has brought this great enterprise to its glorious and beneficent consummation. You have never heard his story; let me give you a few points in it, as a lesson that there is no condition of life in which a man, endowed with native genius, a benevolent spirit, and a courageous patience, may not become a benefactor of nations and of mankind."

After some personal details, which do not concern this history, he speaks of the efforts by which this New-York merchant " brought into being an association of Americans and Englishmen, which contributed from surplus wealth the capital necessary as a basis for the enterprise ;" and then adds :

" It remained to engage the consent and the activity of the Governments of Great Britain and the United States. That was all that remained. Such consent and activity on the part of some one great nation of Europe was all that remained needful for Columbus when he stood ready to bring a new continent forward as a theatre of the world's civilization. But in each case that effort was the most difficult of all. Cyrus W. Field, by assiduity and patience, first secured consent and conditional engagement on the part of Great Britain, and then, less than two years ago, he repaired to Washington. The President and Secretary of State individually favored his proposition ; but the jealousies of parties and sections in Congress forbade them to lend it their official sanction and patronage. He appealed to me. I drew the necessary bill. With the generous aid of others, Northern Representatives and the indispensable aid of the late Thomas J. Rusk, a Senator from Texas, that bill, after a severe contest and long delay, was carried through the Senate of the United States by the majority, if I remember rightly, of one vote, and escaped defeat in the House of Representatives with equal difficulty. I have said the aid of Mr. Rusk was indispensable. If any one has wondered why I, an extreme Northern man, loved and lamented Thomas J. Rusk, an equally extreme Southern man, they have here an explanation. There was no good thing which, as it seemed to me, I could not do in Congress with his aid. When he died, it seemed to me that no good thing could be done by any one. Such was the position of Cyrus W. Field at that stage of the great enterprise. But, thus at last fortified with capital derived from New-York and London, and with the navies of Great Britain and the United States at his command, he has, after trials that would have discouraged any other than a true discoverer, brought the great work to a felicitous consummation. And now the Queen of Great Britain and the President of the United States stand waiting his permission to speak, and ready to speak at his bidding; and the people of these two great countries await only the signal from him to rush into a fraternal embrace which will prove the oblivion of ages of suspicion, of jealousies and of anger."

Mr. Seward may well refer with pride to the part

he took in sustaining this enterprise. He was from the beginning its firmest supporter, as he has been of many other enterprises for the public good. The bill was introduced into the Senate by him, and was carried through mainly by his influence, seconded by Mr. Rusk, Mr. Douglas, and one or two others. It was introduced on the ninth of January, and came up for consideration on the twenty-first. Its friends had hoped that it might pass with entire unanimity. But such was the opposition, that the discussion lasted two days. The report in the pages of the Globe shows that it was a subject of animated and almost angry debate.

This debate is a thing of the past, and cannot be supposed to have a present interest; yet it deserves a brief notice for the reason that it brings out so clearly the objections to aid being given to this enterprise by the Government—objections which are heard even at this late day, and which, as we do not intend to refer to them again, it may be well to notice, once for all, and let them be answered out of the mouths of grave senators.

Probably no measure was ever introduced in Congress for the help of any commercial enterprise, that some member, imagining that it was to benefit a particular section, did not object that it was " unconstitutional!" This objection was well answered in this case by a then honored senator, Mr. Benjamin, of Louisiana, who asked very pertinently: "If we have a right to hire a warehouse at Port Mahon, in the Mediterranean, for storing naval stores, have we not a right to hire a company to carry our messages?" "I should as soon think of questioning the constitutional power of the Government to pay freight to a vessel for carrying its mail-bags across the ocean, as to pay a telegraph company a certain sum per annum for conveying its messages by the use of the electric telegraph."

This touched the precise ground on which the appropriation was asked. In their memorial to the President, the Company had said: "Such a contract will, we suppose, fall within the provisions of the Constitution in regard to postal arrangements, of which this is only a new and improved form."

Mr. Bayard, of Delaware, explained in the same terms the nature of the proposed agreement: "It is a mail operation. It is a Post-Office arrangement. It is for the transmission of intelligence, and that is what I understand to be the function of the Post-Office Department. I hold it, therefore, to be as legitimately within the proper powers of the Government, as the employing of a stage-coach, or a steam-car, or a ship, to transport the mails, either to foreign countries, or to different portions of our own country."

Of course, as in all appropriations of money, the question of expense had to be considered, and here there were not wanting some to cry out against the extravagance of paying seventy thousand dollars a year! We had not then got used to the colossal expenditures of war, and had not grown familiar with the idea of paying three millions a day! Seventy thousand dollars seemed a great sum; but Mr. Bayard in reply reminded them that England then paid nine hundred thousand dollars a year for the transportation of the mails between the United States and England; and argued that it was a very small amount for the great service rendered. He said: "I will venture the assertion, that every Senator on this floor was astonished at the small amount asked for to accomplish this great object. I had supposed it was going to occasion an expense of several hundreds of thousands of dollars a year instead of seventy thousand dollars." "We have sent out ships to make explorations and observations in the Red Sea and in South-America; we sent one or two expensive expeditions to Japan, and published at great cost some elegant books narrating their exploits. The expense even in ships alone, in that instance, was at the rate of twenty to one here, but no cry of economy was then raised." "I look upon this proposition solely as a business measure; in that point of view I believe the Government will obtain more service for the amount of money, than by any other contract that we have ever made, or now can make, for the transmission of intelligence."

As to the expense of furnishing a ship of war to assist in laying the cable, that would be literally nothing. Mr. Douglas well asked: "Will it cost any thing to furnish the use of one of our steamships? They are idle. We have no practical use for them at present. They are in commission. They have their coal on board, and their full armament. They will be rendering no service to us if they are not engaged in this work. If there was nothing more than a question of national pride involved, I would gladly furnish the use of an American ship for that purpose. England tenders one of her national vessels, and why should we not tender one also? It costs England nothing, and it costs us nothing."

Mr. Rusk made the same point, in arguing that ships might be sent to assist in laying the cable, giving this homely but sufficient reason: "I think that is better than to keep them rotting at the navy-yards, with the officers frolicking on shore."

Mr. Douglas urged still further: "But American citizens have commenced this enterprise. The honor and the glory of the achievement, if successful, will be due to American genius and American daring. Why should the American Government be so penurious—I do not know that that is the proper word, for it costs nothing—why should we be actuated by so illiberal a spirit as to refuse the use of one of our steamships to convey the wire when it does not cost one farthing to the Treasury of the United States?"

But behind all these objections of expense and of want of constitutional power, was one greater than them all, and that was ENGLAND! The real animus of the opposition was national jealousy—a fear lest they should be giving some advantage to Great Britain.

This has been always sufficient to excite the hostility of a certain class of politicians. The mention of the name of England has had the same effect on them as a red rag waved before the eyes of a mad bull. No matter what the subject of the proposed coöperation. even if it were purely a scientific expedition, they were sure England was going to profit by it to our injury. So now there were those who felt that in this submarine cable England was literally crawling under the sea to get some advantage of the United States.

This jealousy and hostility spoke loudest from the mouths of Southerners. It is noteworthy that men who, within the last five years, have figured abroad, courting foreign influence against their own country, were then fiercest in denunciation of England. Mason and Slidell voted together against the bill. Butler, of South-Carolina, was very bitter in his opposition—saying, with a sneer, that "this was simply a mail service under the surveillance of Great Britain"—and so was Hunter, of Virginia; while Jones, of Tennessee, bursting with patriotism, found a sufficient reason for his opposition, in that "he did not want any thing to do with England or Englishmen!"

But it should be said in justice, that to this general hostility of the South there were some exceptions. Benjamin, of Louisiana, gave the bill an earnest support; so did Mallory, of Florida, Chairman of the Naval Committee; and especially that noble Southerner, Rusk, of Texas, " with whose aid," as Mr. Seward said, " it seemed that there was no good thing which he could not do in Congress." Mr. Rusk declared that he regarded it as "the great enterprise of the age," and expressed his surprise at the very moderate subsidy asked for, only seventy thousand dollars a year, and declared that, " with a reasonable prospect of success in an enterprise of this description, calculated to produce such beneficial results, he should be willing to vote two hundred thousand dollars."

But with the majority of Southern Senators, there was a repugnance to acting in concert with England, which could not be overcome. They argued that this was not truly a line between England and the United

States, but between England and her own colonies— a line of which she alone was to reap the benefit. *Both its termini were in the British possessions.* In the event of war this would give a tremendous advantage to the power holding both ends of the line. All the speakers harped on this string; and it may be worth a page or two to see how this was met and answered. Thus spoke Mr. Hunter, of Virginia:

"There is another matter which seems to me to require some safeguards. Both the termini of this telegraph line are in the British dominions. What security are we to have that in time of war we shall have the use of the telegraph as well as the British Government?"

The answer of Mr. Seward may satisfy our English friends that he is not animated by any violent hostility to that country:

"It appears not to have been contemplated by the British Government that there would ever be any interruption of the amicable relations between the two countries. Therefore nothing was proposed in their contract for the contingency of war.

"That the two termini are both in the British dominions is true; but it is equally true that there is no other terminus on this continent where it is practicable to make that communication except in the British dominions. We have no dominions on the other side of the Atlantic Ocean. There is no other route known on which the telegraphic wire could be drawn through the ocean so as to find a proper resting place or anchorage except this. The distance on this route is seventeen hundred miles. It is not even known that the telegraphic wire will carry the fluid with sufficient strength to communicate across those seventeen hundred miles. That is yet a scientific experiment, and the Company are prepared to make it.

"In regard to war, all the danger is this: There is a hazard of war at some future time, and whatever arrangements we might make, war would break them up — at least, war would probably break them up. There can be no stipulation of treaty that would save us the benefit desired. My own hope is, that after the telegraphic wire is once laid, there will be no more war between the United States and Great Britain. I believe that whenever such a connection as this shall be made, we diminish the chances of war, and diminish them in such a degree, that it is not necessary to take them into consideration at the present moment.

"Let us see where we are? What shall we gain by refusing to enter into this agreement? If we do not make it, the British Government has only to add ten thousand pounds sterling more annually, and they have the whole monopoly of this wire, without any stipulation whatever—not only in war but in peace. If we make this contract with the Company, we at least secure the benefit of it in time of peace, and we postpone and delay the dangers of war. If there shall ever be war, it would abrogate all treaties that can be made in regard to this subject, unless it be true, as the honorable Senator from Virginia thinks, that treaties can be made which will be regarded as obligatory by nations in time of war. If so, we have all the advantages in time of peace, for the purpose of making such treaties hereafter, without the least reason to infer that there would be any reluctance on the part of the British Government to enter into that negotiation with us, if we should desire to do so. The British Government, if it had such a disposition as the honorable Senator supposes, would certainly have proposed to monopolize all this telegraphic line, instead of proposing to divide it." *

* It is worthy of notice, that when the Bill granting a charter to the Atlantic Telegraph Company was offered in the British Parliament, at least one nobleman found fault with it on this very ground, that it gave away important advantages which properly belonged to England, and which she ought to reserve to herself:

"In the House of Lords, on the twentieth of July, 1857, on the motion for the third reading of the Telegraph Company's bill,

"Lord Redesdale called attention to the fact that, although the termini of the proposed telegraph were both in her Majesty's dominions, namely, in Ireland and Newfoundland, the American Government were

Mr. Hale spoke in the same strain:

"It seems to me that the war spirit and the contingencies of war are brought in a little too often upon matters of legislation which have no necessary connection with them. If we are to be governed by considerations of that sort, they would paralyze all improvement; they would stop the great appropriations for commerce; they would at once neutralize that policy which sets our ocean steamers afloat. Nobody pretends that the intercourse which is kept up between Great Britain and this country by our ocean steamers would be continued in time of war; nor the communication with France or other nations.

to enjoy the same priority as the British Government with regard to the transmission of messages. It was said that this equal right was owing to the fact that a joint guarantee had been given by the two Governments. *He thought, however, it would have been far better policy on the part of her Majesty's Government if they had either undertaken the whole guarantee themselves, and thus had obtained free and sole control over the connecting line of telegraph, or had invited our own colonies to participate in that guarantee, rather than have allowed a foreign government to join in making it.* At the same time, if the clause in question had the sanction of her Majesty's ministry, it was not his intention to object to it.

"Earl Granville said this telegraph was intended to connect two great countries, and, as the two Governments had gone hand in hand with regard to the guarantee, it seemed only reasonable that both should have the same rights as to transmitting messages.

"The bill was then read a third time and passed."

"If we are deterred for that reason, we shall be pursuing a policy that will paralyze improvements on those parts of the coast which lie contiguous to the lakes. The city of Detroit will have to be abandoned, beautiful and progressive as it is, because in time of war the mansions of her citizens there lie within the range of British guns.

"What will the suspension bridge at Niagara be good for in a time of war? If the British cut off their end of it, our end will not be worth much. I believe that among the things which will bind us together in peace, this telegraphic wire will be one of the most potent. It will bind the two countries together literally with cords of iron that will hold us in the bonds of peace. I repudiate entirely the policy which refuses to adopt it, because in time of war it may be interrupted. Such a policy as that would drive us back to a state of barbarism. It would destroy the spirit of progress; it would retard improvement; it would paralyze all the advances which are making us a more civilized, and a more informed and a better people than the one which preceded us."

Mr. Douglas cut the matter short by saying:

"I am willing to vote for this bill as a peace measure, as a commercial measure—but not as a war measure; and when war comes, let us rely on our power and ability to take this end of the wire, and keep it."

Mr. Rusk said: "The advantages of this work will be mutual, and must be mutual, between the United States and Great Britain. It is impossible for one nation at this age to get a great advantage over another in means of communication, because when a communication is made, it will be open to the intelligence and enterprise and capital of all."

Mr. Benjamin said: "The sum of money that this Government proposes to give for the use of this telegraph will amount, in the twenty five years, to something between £300,000 and £400,000. Now, if this be a matter of such immense importance to Great Britain—if this be the golden opportunity—and if, indeed, her control of this line be such a powerful engine, whether in war or in peace, is it not most extraordinary that she proposes to us a full share in its benefits and in its control, and allows to our Government equal rights with herself in the transmission of communications for the sum of about £300,000, to be paid in annual instalments through twenty-five years? If this be, indeed, a very important instrumentality in behalf of Great Britain for the conduct of her commerce, the government of her possessions, or the efficient action of her troops in time of war, the £300,000 expended upon it are but as a drop in the bucket when compared with the immense resources of that empire. I think, therefore, we may as well discard from our consideration of this subject all these visions about the immense importance of the governmental aid in this matter, to be rendered under the provisions of this bill.

"Mr. President, let us not always be thinking of war; let us be using means to preserve peace. The amount that would be expended by this Government in six months' war with Great Britain, would far exceed every thing that we shall have to pay for the use of this telegraphic line for the entire twenty-five years of the contract; and do you not believe that this instrumentality will be sufficiently efficient to bind together the peace, the commerce, and the interests of the two countries, so as even to defer a war for six months or twelve months, if one should ever become inevitable, beyond the period at which it would otherwise occur? If it does that, it will in six or eight or nine months repay the expenditures of twenty-five years.

"Again, sir, I say, if Great Britain wants it for war, she will put it there at her own expense. It is not three hundred thousand pounds, or four hundred thousand pounds, that will arrest her. If, on the contrary, this be useful to commerce—useful in an eminent degree—useful for the preservation of peace, then I confess I feel some pride that my country should aid in establishing it. I confess I feel a glow of something like pride that I belong to the great human family when I see these triumphs of science, by which mind is brought into instant communication with mind across the intervening oceans, which, to our unenlightened forefathers, seemed placed there by Providence as an eternal barrier to communication between man and man. Now, sir, we speak from minute to minute. Scarcely can a gun be fired in war on the European shore ere its echoes will reverberate among our own mountains, and be heard by every citizen in the land. All this is a triumph of science—of American genius, and I for one feel proud of it, and feel desirous of sustaining and promoting it."

Mr. Douglas said: "Our policy is essentially a policy of peace. We want peace with the whole world, above all other considerations. There never has been a time in the history of this Republic, when peace was more essential to our prosperity, to our advancement, and to our progress, than it is now. We have made great progress in time of peace—an almost inconceivable progress since the last war with Great Britain. Twenty-five years more of peace will put us far in advance of any other nation on earth."

It was fit that Mr. Seward, who introduced the bill, and opened the debate, should close in words that now seem prophetic, and show the large wisdom, looking before and after, of this eminent statesman:

"There was an American citizen who, in the year 1770, or thereabout, indicated to this country, to Great Britain, and to the world, the use of the lightning for the purposes of communication of intelligence, and that was Dr. Franklin. I am sure that there is not only no member of the Senate, but no American citizen, however humble, who would be willing to have struck out from the achievements of American invention this great discovery of the lightning as an agent for the uses of human society.

"The suggestion made by that distinguished and illustrious American was followed up some fifty years afterward by another suggestion and another indication from another American, and that was Mr. Samuel F. B. Morse, who indicated to the American Government the means by which the lightning could be made to write, and by which the telegraphic wires could be made to supply the place of wind and steam for carrying intelligence.

"We have followed out these suggestions of these eminent Americans hitherto, and I am sure at a very small cost. The Government of the United States appropriated $40,000 to test the practicability of Morse's suggestion; the $40,000 thus expended established its practicability and its use. Now, there is no person on the face of the globe who can measure the price at which, if a reasonable man, he would be willing to strike from the world the use of the magnetic telegraph as a means of communication between different portions of the same country. This great invention is now to be brought into its further, wider, and broader use—the use by the general society of nations, international use, the use of the society of mankind. Its benefits are large—just in proportion to the extent and scope of its operation. They are not merely benefits to the Government, but they are benefits to the citizens and subjects of all nations and of all States.

"I might enlarge further on this subject, but I forbear to do so, because I know that at some future time I shall come across the record of what I have said to-day. I know that *then what I have said to-day, by way of anticipation, will fall so far short of the reality* of benefits which individuals, States, and nations will have derived from this great enterprise, that I shall not reflect upon it without disappointment and mortification."

After such arguments, it should seem that there could be but one opinion, and yet the bill passed the Senate by only *one* majority! It also had to run the gauntlet of the House of Representatives, where it encountered the same hostility. But at length it got through, and was signed by President Pierce on the third of March, the day before he went out of office. Thus it became a law.

CHAPTER VIII.

RETURN TO ENGLAND. THE NIAGARA—CAPTAIN HUDSON. THE AGAMEMNON. EXPEDITION OF 1857. SAILING FROM IRELAND. SPEECH OF THE EARL OF CARLISLE. THE CABLE BROKEN.

SCARCELY was the business with the American Government completed, before Mr. Field was recalled to England. Once more upon the waves, he forgot the long delay and the vexatious opposition which he left behind—the fogs of Newfoundland, and the denser fogs of Washington. He was bound for England, and there at least the work did not stand still. All winter long the wheels of the machinery had kept in motion. The cable was uncoiling its mighty folds to a length sufficient to span the Atlantic, and at last there was hope of victory.

Although the United States Government had seemed a little ungracious in its delay, it yet rendered, this year and next, important service to the Atlantic Telegraph. Already it had prepared the way for it, by the deep-sea soundings, which it was the first to take across the Atlantic. It now rendered additional and substantial aid in lending to this enterprise the two finest ships in the American navy—the Niagara and the Susquehanna. The former especially deserves notice, as she has become a historical ship. She was built some dozen years ago by the lamented George Steers—a name celebrated among our marine architects as the constructor of the famous yacht America, that " race-horse of the sea," which had crossed the Atlantic, and carried off the prize in the British Channel from the yachts of England—and was designed to be a model of naval architecture. She was the largest steam-frigate in the world, exceeding in tonnage the heaviest line-of-battle ship in the English navy, and yet so finely modelled that, propelled only by a screw, she could easily make ten or twelve miles an hour. Notwithstanding her huge bulk, she was intended to carry but twelve guns—being one of the first ships in our navy to substitute a few enormous Dahlgrens for half a dozen times as many fifty-six-pounders. This was the beginning of that revolution in naval warfare, which has been carried to such extent in the Monitors and other ironclads introduced in our recent civil war. Each gun weighed fourteen tons—requiring a crew of twenty-five men to wield it—and threw a shell of one hundred and thirty pounds a distance of three miles. One or two broadsides from such a deck would sink an old-fashioned seventy-four, or even a ninety or hundred-gun ship.

But as the Niagara was now to go on an errand of peace, this formidable armament was not taken on board. She was built with what is known as a "flush deck," clear from stem to stern, and being without her guns, was left free for the more peaceful burden that she was to bear. When the orders were received from Washington, she was lying at the Brooklyn Navy-Yard, but began immediately to prepare for her expedition. Bulkheads were knocked down, above and below, to make room for that huge serpent—longer a thousand times than any fabled monster of the deep—that was to be coiled within her sides. These preparations occupied four or five weeks. On the twenty-second of April, she made a trial trip down the bay, and two days after sailed for England, in command of Captain William L. Hudson, one of the oldest and best officers in our navy, who, to his past services to his country, was now to add another in the expeditions of this and the following year. He had with him as Chief Engineer Mr. William E. Everett, whose mechanical genius proved so important in constructing the paying-out machinery which finally laid the cable.

Besides the regular ship's crew, no one was received on board except two officers of the Russian navy— Captain Schwartz and Lieutenant Kolobnin — who were permitted by our Government, as an act of national courtesy, to go out to witness this great experiment; and Professor Morse, who went as the electrician of the Newfoundland Company. The regulations of the navy did not admit correspondents of the press; but Professor Morse was permitted to take a secretary, and chose Mr. Mullaly, who had thus an opportunity to witness all the preparations on land and sea, and to furnish those minute and detailed accounts of the several expeditions, which contribute some important chapters in the history of this enterprise.

The Niagara arrived out on the fourteenth of May, and cast anchor off Gravesend, about twenty-five miles below London. As it was the first time—at least for many years—that an American ship of war had appeared in the Thames, this fact, with her huge proportions and the object for which she came, attracted a crowd of visitors. Every day, from morning to night, a fleet of boats was around her, and men and women thronged over her sides. Every body was welcome. All were received with the utmost courtesy, and allowed access to all parts of the ship. Among these were many visitors of distinction. Here came Lady Franklin to thank the generous nation that had sent two expeditions to recover her husband lost amid Polar seas. She was, of course, the object of universal attention and respectful sympathy.

While lying in the Thames, the Agamemnon, that was to take the other half of the cable, passed up the river. This, too, was a historical ship, having borne the flag of the British admiral at the bombardment of Sebastopol, and distinguished herself by steaming up within a few hundred yards of the guns of the fortress. After passing through the fires of that terrible day, she was justly an object of pride to Britons, whose hearts swelled as they saw this oak-ribbed leviathan, that had come " out of the gates of death, out of the jaws of hell," now preparing to take part in achievements of peace, not less glorious than those of war. She was under command of Captain Noddal, of the Royal Navy.

As the Agamemnon came up the river in grand style, she recognized the Niagara lying off Gravesend, and manning her yards, gave her a succession of those English hurras so stirring to the blood, when heard on land or sea, to which our tars replied with lusty American cheers. It was pleasant to observe, from this time, the hearty good-will that existed between the officers and crews of the two ships, who in their exertions for the common object, were animated only by a generous rivalry.

A few days after, the Niagara was joined by the Susquehanna, Captain Sands, which had been ordered from the Mediterranean to take part also in the expedition. She was a fit companion ship, being the largest side-wheel steamer in our navy, as the other was the largest propeller. Both together, they were worthy representatives of the American navy.

When the Niagara arrived in the Thames, it was supposed she would take on board her half of the cable from the manufactory of Glass, Elliot & Co., at Greenwich; but on account of her great length, it was difficult to bring her up alongside the wharf in front of the works. This was therefore left to the Agamemnon, while the Niagara was ordered around to Liverpool, to take the other half from the works of Newall & Co., at Birkenhead, opposite that city. Accordingly she left Gravesend on the fifth of June, and reached Portsmouth the next day, where she remained a fortnight, to have some further alterations to fit her to receive the cable. Although she had been already pretty well "scooped out," fore and aft, the cry was still for room. Officers had to shift for themselves, as their quarters were swept away to make a wider berth for their iron guest. But all submitted with excellent grace. Like true sailors, they took it "gaily" as if they were only clearing the decks to go into battle. Among other alterations for safety, was a framework or "cage" of iron, which was put over the stern of the ship, to keep the cable from getting entangled in the screw. As soon as these were completed, the Niagara left for Liverpool, and on the twenty-second of June cast anchor in the Mersey. Here she attracted as much attention as in the Thames, being crowded with visitors during the week; and on Sundays, when none were received on board, the river-boats sought to gratify public curiosity by sailing around her. The officers of the ship at the same time were objects of constant hospitality, both from private citizens and from the public authorities. The Mayor of Liverpool gave them a dinner, the Chamber of Commerce another, while the Americans in Liverpool entertained them at a grand banquet on the fourth of July—the first public celebration of our national anniversary ever had in that city.

But while these festivities were kept up on shore, hard work was done on board the ship. To coil thirteen hundred miles of cable was an immense undertaking. Yet it was all done by the sailors themselves. No compulsion was used, and none was needed. No sooner was there a call for volunteers, than men stepped forward in greater numbers than could be employed. Out of these were chosen one hundred and twenty stalwart fellows, who were divided into two gangs of sixty men, and each gang into watches of thirty, which relieved each other, and all went to work with such enthusiasm, that in three weeks the herculean task was completed. The event was celebrated by a final dinner given by the shareholders of the Atlantic Telegraph Company residing at Liverpool to Captain Hudson, and Captain Sands of the Susquehanna, whose arrival in the Mersey enabled them to extend their hospitalities to the officers of both ships.

While the Niagara was thus doing her part, the same scene was going on on board the Agamemnon, which was still lying in the Thames. There the work was completed about the same day, and the occasion duly honored by a scene as unique as it was beautiful. Says the London Times of July twenty-fourth:

"All the details connected with the manufacture and stowage of the cable are now completed, and the conclusion of the arduous labor was celebrated yesterday with high festivity and rejoicing. All the artisans who have been engaged upon the great work, with their wives and families, a large party of the officers, with the sailors from the Agamemnon, and a number of distinguished scientific visitors, were entertained upon this occasion at a kind of *fête champêtre* at Belvidere house, the seat of Sir Culling Eardley, near Erith. The festival was held in the beautiful park which had been obligingly opened by Sir Culling Eardley for the purpose. Although in no way personally interested in the project, the honorable baronet has all along evinced the liveliest sympathy with the undertaking, and himself proposed to have the completion of the work celebrated in his picturesque grounds. The manufacturers, fired with generous emulation, erected spacious tents on the lawn, and provided a magnificent banquet for the guests, and a substantial one for the sailors of the Agamemnon and the artificers who had been employed in the construction of the cable. By an admirable arrangement, the guests were accommodated at a vast semi-circular table, which ran round the whole pavilion, while the sailors and workmen sat at a number of long tables arranged at right angles with the chord, so that the general effect was that all dined together, while at the same time sufficient distinction was preserved to satisfy the most fastidious. The three centre tables were occupied by the crew of the Agamemnon, a fine, active body of young men, who paid the greatest attention to the speeches, and drank all the toasts with an admirable punctuality, at least so long as their three pints of beer per man lasted; but we regret to add that, what with the heat of the day and the enthusiasm of Jack in the cause of science, the mugs were all empty long before the chairman's list of toasts had been gone through. Next in interest to the sailors were the workmen and their wives and babies, all being permitted to assist at the great occasion. The latter, it is true, sometimes squalled at an affecting peroration, but that rather improved the effect than otherwise, and the presence of these little ones only marked the genuine good feeling of the employers, who had thus invited not only their workmen but their workmen's families to the feast. It was a momentary return to the old patriarchal times, and every one present seemed delighted with the experiment."

Speeches were made by Sir Culling Eardley, by Mr. Cardwell, of the House of Commons, Mr. Brooking, one of the Directors, by Professor Morse, and others. Mr. Field read a letter from President Buchanan, saying that he should feel honored if the first message should be one from Queen Victoria to himself, and that he " would endeavor to answer it in a spirit and manner becoming a great occasion."

Thus labor and feasting being ended, the Niagara and the Susquehanna left Liverpool the latter part of July, and steamed down St. George's Channel to Queenstown, which was to be the rendezvous of the telegraphic squadron, and here they were joined by the Agamemnon and the Leopard, which was to be her consort. The former, as she entered the harbor, came to anchor about a third of a mile from the Niagara. The presence of the two ships which had the cable on board, gave an opportunity which the electricians had desired to test its integrity. Accordingly one end of each cable was carried to the opposite ship, and so joined as to form a continuous length of twenty-five hundred miles, both ends of which were on board the Agamemnon. One end was then connected with the apparatus for transmitting the electric current, and on a sensitive galvanometer being attached to the other end, the whole cable was tested from end to end, and found to be perfect. These experiments were continued for two days with the same result. This inspired fresh hopes for the success of the expedition, and in high spirits they bore away for the harbor of Valentia.

It had been for some time a matter of discussion, where they should begin to lay the cable, whether from the coast of Ireland, or in mid-ocean, the two ships making the junction there, and dropping it to the bottom of the sea, and then parting, one to the east and the other to the west, till they landed their ends on the opposite shores of the Atlantic. This was the plan adopted the following year, and which finally proved successful. It was the one preferred by the engineers now, but the electricians favored the other course, and their counsel prevailed. It was therefore decided *to submerge the whole cable in a continuous line from Valentia Bay to Newfoundland.* The Niagara was to lay the first half from Ireland to the middle of the Atlantic; the end would then be joined to the other half on board the Agamemnon, which would take it on to the coast of Newfoundland. During the whole process the four vessels were to remain together and give whatever assistance was required. While it was being laid down, messages were to be sent back to Valentia, reporting each day's progress.

As might be supposed, the mustering of such a fleet of ships, and the busy note of preparation which had been heard for weeks, produced a great sensation in this remote part of Ireland. The people from far and near, gathered on the hills and looked on in silent wonder.

To add to the dignity of the occasion, the Lord Lieutenant came down from Dublin to witness the departure of the expedition. Happily this was a nobleman well fitted to represent his own country, and to command audience from ours. The Earl of Carlisle—better known among us as Lord Morpeth—had travelled in the United States a few years before, and shown himself one of the most intelligent and liberal foreigners that ever visited America. No representative of England could on that day have stood upon the shores of Ireland, and stretched out his hand to his kindred beyond the sea with more assurance that his greeting would be warmly responded to. And never did one speak more aptly words of wisdom and of peace. We read them still with admiration for their beauty and their eloquence, and with an interest more tender but more sad, that this great and good man—the true friend of his own country and of ours—has gone to his grave. To quote his own words is the best tribute to his memory, and will do more than any eulogy we can pay to keep it fresh and green in the hearts of Americans. On his arrival at Valentia, he was entertained by the Knight of Kerry at one of those public breakfasts so much in fashion in England, at which in response to a toast in his honor, after making his personal acknowledgments, he said:

"I believe, as your worthy chairman has already hinted, that I am probably the first Lieutenant of Ireland who ever appeared upon this lovely strand. At all events, no Lord Lieutenant could have come amongst you on an occasion like the present. Amidst all the pride and the stirring hopes which cluster around the work of this week, we ought still to remember that we must speak with the modesty of those who begin and not of those who close an experiment, and it behooves us to remember that the pathway to great achievements has frequently to be hewn out amidst risks and difficulties, and that preliminary failure is even the law and condition of the ultimate success. [Loud cheers.] Therefore, whatever disappointments may possibly be in store, I must yet insinuate to you that in a cause like this it would be criminal to feel discouragement. [Cheers.] In the very design and endeavor to establish the Atlantic Telegraph there is almost enough of glory. It is true if it be only an attempt there would not be quite enough of profit. I hope that will come, too; but there is enough of public spirit, of love for science, for our country, for the human race, almost to suffice in themselves. However, upon this rocky frontlet of Ireland, at all events, to-day we will presume upon success. We are about, either by this sundown or by to-morrow's dawn, to establish a new material link between the Old World and the New. Moral links there have been—links of race, links of commerce, links of friendship, links of literature, links of glory; but this, our new link, instead of superseding and supplanting the old ones, is to give a life and an intensity which they never had before. [Loud cheers.] Highly as I value the reputations of those who have conceived, and those who have contributed to carry out this bright design—and I wish that so many of them had not been unavoidably prevented from being amongst us at this moment*— highly as I estimate their reputation, yet I do not compliment them with the idea that they are to efface or dim the glory of that Columbus, who, when the large vessels in the harbor of Cork yesterday weighed

---

* All of the ships had not arrived.

their anchors, did so on that very day three hundred and sixty-five years ago—it would have been called in Hebrew writ a year of years—and set sail upon his glorious enterprise of discovery. They, I say, will not dim or efface his glory, but they are now giving the last finish and consummation to his work. Hitherto the inhabitants of the two worlds have associated perhaps in the chilling atmosphere of distance with each other—a sort of bowing distance; but now we can be hand to hand, grasp to grasp, pulse to pulse. [Cheers.] The link, which is now to connect us, like the insect in the immortal couplet of our poet:

While exquisitely fine,
Feels at each thread and lives along the line.

And we may feel, gentlemen of Ireland, of England, and of America, that we may take our stand here upon the extreme rocky edge of our beloved Ireland; we may, as it were, leave in our rear behind us the wars, the strifes, and the bloodshed of the elder Europe, and I fear I may say, of the elder Asia; and we may pledge ourselves, weak as our agency may be, imperfect as our powers may be, inadequate in strict diplomatic form as our credentials may be, yet, in the face of the unparalleled circumstances, of the place and the hour, in the immediate neighborhood of the mighty vessels whose appearance may be beautiful upon the waters, even as are the feet upon the mountains of those who preach the Gospel of peace—as an homage due to that serene science which often affords higher and holier lessons of harmony and good will than the wayward passions of man are always apt to learn—in the face and in the strength of such circumstances, let us pledge ourselves to eternal peace between the Old World and the New."

While these greetings were exchanged on shore, only the smaller vessels of the squadron had arrived. But in a few hours the great hulls of the Niagara and the Agamemnon, followed by the Leopard and the Susquehanna, were seen in the horizon, and soon they all cast anchor in the bay. As the sun went down in the west, shining still on that other hemisphere which they were going to seek, its last rays fell on an expedition more suggestive and hopeful than any since that of Columbus from the shores of Spain, and upon navigators not unworthy to be called his followers.

The whole squadron was now assembled, and made quite a naval array. There were present in the little harbor of Valentia seven ships—the stately Niagara, which was to lay the half of the cable from Ireland, and her consort, the Susquehanna, riding by her side; while floating the flag of England, were the Agamemnon, which was appointed to lay the cable on the American side, and her consort, the Leopard. Beside these high-decked ships of war, the steamer Advice had come round to give, not merely advice but lusty help in landing the cable at Valentia; and the little steamer Willing Mind, with a zeal worthy of her name, was flying back and forth between ship and shore, lending a hand wherever there was work to be done; and the Cyclops, under the experienced command of Captain Dayman, who had made the deep-sea soundings across the Atlantic only the month before, here joined the squadron to lead the way across the deep. This made five English ships, with but two American; but to keep up our part, there were two more steamers on the other side of the sea, the Arctic, under Lieutenant Berryman, and the Company's steamer Victoria, to watch for the coming of the fleet off the coast of Newfoundland, and help in landing the cable on the shores of the New World.

It was now Tuesday evening, the fourth of August, too late to undertake the landing that night, but preparations were at once begun for it the next morning. Says the correspondent of the Liverpool Post:

"The ships were visited in the course of the evening by the Directors and others interested in the great undertaking, and arrangements were immediately commenced on board the Niagara for paying out the shore rope for conveyance to the mainland. These arrangements were fully perfected by Wednesday morning;

but for some hours the state of the weather rendered it doubtful whether operations could be safely proceeded with. Toward the afternoon the breeze calmed down, and at two o'clock it was decided that an effort should be made to land the cable at once. The process of uncoiling into the small boats commenced at half-past two, and the scene at this period was grand and exciting in the highest degree.

"Valentia Bay was studded with innumerable small craft, decked with the gayest bunting—small boats flitted hither and thither, their occupants cheering enthusiastically as the work successfully progressed. The cable boats were managed by the sailors of the Niagara and Susquehanna, and it was a well-designed compliment, and indicative of the future fraternization of the nations, that the shore rope was arranged to be presented at this side the Atlantic to the representative of the Queen, by the officers and men of the United States navy, and that at the other side British officers and sailors should make a similar presentation to the President of the great Republic.

"From the main land the operations were watched with intense interest. For several hours the Lord Lieutenant stood on the beach, surrounded by his staff and the directors of the railway and telegraph companies, waiting the arrival of the cable, and when at length the American sailors jumped through the surge with the hawser to which it was attached, his Excellency was among the first to lay hold of it and pull it lustily to the shore. Indeed every one present seemed desirous of having a hand in the great work; and never before perhaps were there so many willing assistants, at 'the long pull, the strong pull, and the pull altogether.'

"At half-past seven o'clock the cable was hauled on shore, and formal presentation was made of it to the Lord Lieutenant by Captain Pennock, of the Niagara; his Excellency expressing a hope that the work so well begun would be carried to a satisfactory completion."

The wire having been secured to a house on the beach, the Reverend Mr. Day, of Kenmore, advanced and offered the following prayer:

"O Eternal Lord God, who alone spreadest out the heavens, and rulest the raging of the sea; who hast compassed the waters with bounds, till day and night come to an end; and whom the winds and the sea obey; look down in mercy, we beseech thee, upon us thy servants, who now approach the throne of grace; and let our prayer ascend before thee with acceptance. Thou hast commanded and encouraged us, in all our ways, to acknowledge thee, and to commit our works to thee; and thou hast graciously promised to direct our paths, and to prosper our handiwork. We desire now to look up to thee; and believing that without thy help and blessing, nothing can prosper or succeed, we humbly commit this work, and all who are engaged in it, to thy care and guidance. Let it please thee to grant to us thy servants wisdom and power, to complete what we have been led by thy Providence to undertake; that being begun and carried on in the spirit of prayer, and in dependence upon thee, it may tend to thy glory: and to the good of all nations, by promoting the increase of unity, peace, and concord.

"Overrule, we pray thee, every obstacle, and remove every difficulty which would prevent us from succeeding in this important undertaking. Control the winds and the sea by thy Almighty power, and grant us such favorable weather that we may be enabled to lay the Cable safely and effectually. And may·thy hand of power and mercy be so acknowledged by all, that the language of every heart may be, 'Not unto us, O Lord, not unto us, but unto thy name give glory,' that so thy name may be hallowed and magnified in us and by us.

"Finally, we beseech thee to implant within us a spirit of humanity and childlike dependence upon thee; and teach us to feel as well as to say, 'If the Lord will, we shall do this or that.'

"Hear us, O Lord, and answer us in these our petitions, according to thy precious promise for Jesus Christ's sake. Amen."

The Lord Lieutenant then spoke once more—words that amid such a scene and at such an hour, sank into all hearts :

"My American, English, and Irish friends, I feel at such a moment as this that no language of mine can be becoming except that of prayer and praise. However, it is allowable to any human lips, though they have not been specially qualified for the office, to raise the ascription of 'Glory to God in the highest; on earth peace, good-will to men.' That, I believe, is the spirit in which this great work has been undertaken; and it is this reflection that encourages me to feel confident hopes in its final success. I believe that the great work now so happily begun will accomplish many great and noble purposes of trade, of national policy, and of empire. But there is only one view in which I will present it to those whom I have the pleasure to address. You are aware—you must know, some of you, from your own experience—that many of your dear friends and near relatives have left their native land to receive hospitable shelter in America. Well, then, I do not expect that all of you can understand the wondrous mechanism by which this great undertaking is to be carried on. But this, I think, you all of you understand. If you wished to communicate some piece of intelligence straightway to your relatives across the wide world of waters—if you wished to tell those whom you know it would interest in their heart of hearts, of a birth, or a marriage, or, alas, a death, among you, the little cord, which we have now hauled up to the shore, will impart that tidings quicker than the flash of the lightning. Let us indeed hope, let us pray that the hopes of those who have set on foot this great design, may be rewarded by its entire success; and let us hope, further, that this Atlantic Cable will, in all future time, serve as an emblem of that strong cord of love which I trust will always unite the British islands to the great continent of America. And you will join me in my fervent wish that the Giver of all Good, who has enabled some of his servants to discern so much of the working of the mighty laws by which he fills the universe, will further so bless this wonderful work, as to make it even more to serve the high purpose of the good of man, and tend to his great glory. And now, all my friends, as there can be no project or undertaking which ought not to receive the approbation and applause of the people, will you join with me in giving three hearty cheers for it? [Loud cheering.] Three cheers are not enough for me—they are what we give on common occasions—and as it is for the success of the Atlantic Telegraph Cable, I must have at least one dozen cheers. [Loud and protracted cheering.]"

Mr. Brooking, the Chairman of the Executive Committee of the Atlantic Telegraph Company, then expressed the thanks which all felt, to the Lord Lieutenant, for his presence on that occasion.

Then there were loud calls for Mr. Field. He answered in a few short sentences:

"I have no words to express the feelings which fill my heart to-night—it beats with love and affection for every man, woman and child who hears me. I may say, however, that, if ever at the other side of the waters now before us, any one of you shall present yourselves at my door and say that you took hand or part, even by an approving smile, in our work here to-day, you shall have a true American welcome. I cannot bind myself to more, and shall merely say: 'What God has joined together, let not man put asunder.'"

Thus closed this most interesting scene. The Lord Lieutenant was obliged to return at once to the capital. He therefore left, and posted that night to Killarney, and the next day returned by special train to Dublin, leaving the ships to complete the work so happily begun.

The landing of the cable took place on Wednesday, the fifth of August, near the hour of sunset. As it was too late to proceed that evening, the ships remained at anchor till the morning. They got under weigh at an early hour, but were soon checked by an accident which detained them another day. Before they had gone five miles, the heavy shore end of the cable caught in the machinery and parted. The Niagara put back, and the cable was "underrun" the whole distance. At length the end was lifted out of the water and spliced to the gigantic coil, and as it dropped safely to the bottom of the sea, the mighty ship began to stir. At first she moved very slowly, not more than two miles an hour, to avoid the danger of accident; but the feeling that they are at last away is itself a relief. The ships are all in sight, and so near that they can hear each other's bells. The Niagara, as if knowing that she is bound for the land out of whose forests she came, bends her head to the waves, as her prow is turned toward her native shores.

Slow passed the hours of that day. But all went well, and the ships were moving out into the broad Atlantic. At length the sun went down in the west, and stars came out on the face of the deep. But no man slept. A thousand eyes were watching a great experiment as those who have a personal interest in the issue. All through that night, and through the anxious days and nights that followed, there was a feeling in every soul on board, as if a friend in the cabin were at the turning-point of life or death, and they were watching beside him. There was a strange, unnatural silence in the ship. Men paced the deck with soft and muffled tread, speaking only in whispers, as if a loud voice or a heavy footfall might snap the vital chord. So much had they grown to feel for the enterprise, that the cable seemed to them like a human creature, on whose fate they hung, as if it were to decide their own destiny.

There are some who will never forget that first night at sea. Perhaps the reäction from the excitement on shore made the impression the deeper. What strange thoughts came to them as they stood on the deck and watched that mysterious chord disappearing in the darkness, and gliding to its ocean bed! There are certain moments in life when every thing comes back upon us, when the events of years seem crowded into an hour. What memories came up in those long night hours! How many on board that ship thought of homes beyond the sea, of absent ones, of the distant and the dead! Such thoughts, mingling with those suggested by the scene around, added to the solemnity of the hour, and left an impression which can never be forgotten.

But with the work in hand all is going on well. There are vigilant eyes on deck. Mr. Bright, the engineer of the Company, is there, and Mr. Everett, Mr. De Sauty, the electrician, and Professor Morse. The paying-out machinery does its work, and though it makes a constant rumble in the ship, that dull, heavy sound is music to their ears, as it tells them that all is well. If one should drop to sleep, and wake up at night, he has only to hear the sound of " the old coffee-mill," and his fears are relieved, and he goes to sleep again.

Saturday was a day of beautiful weather. The ships were getting farther away from land, and began to steam ahead at the rate of four and five miles an hour. The cable was paid out at a speed a little faster than that of the ship, to allow for any inequalities of surface on the bottom of the sea. While it was thus going overboard, communication was kept up constantly with the land. Every moment the current was passing between ship and shore. The communication was as perfect as between Liverpool and London, or Boston and New-York. Not only did the electricians telegraph back to Valentia the progress they were making, but the officers on board sent messages to their friends in America, to go out by the steamers from Liverpool. The heavens seemed to smile on them that day. The coils came up from below the deck without a kink, and unwinding themselves easily, passed over the stern into the sea. Once or twice an alarm was created by the cable being thrown off the wheels. This was owing to the sheaves not being wide enough and deep enough, and being filled with tar, which hardened in the air. This was a great defect of the machinery which was remedied in the later expeditions. Still it worked well, and so long as those terrible brakes kept off their iron gripe, it might work through to the end.

All day Sunday the same favoring fortune continued; and when the officers, who could be spared from the deck, met in the cabin, and Captain Hudson read the service, it was with subdued voices and grateful hearts they responded to the prayers to Him who spreadeth out the heavens, and ruleth the raging of the sea.

On Monday they were over two hundred miles to sea. They had got far beyond the shallow waters off the coast. They had passed over the submarine mountain which figures on the charts of Dayman and Berryman, and where Mr. Bright's log gives a descent from five hundred and fifty to seventeen hundred and fifty fathoms within eight miles! Then they came to the deeper waters of the Atlantic, where the cable sank to the awful depth of two thousand fathoms. Still the iron cord buried itself in the waves, and every instant the flash of light in the darkened telegraph room told of the passage of the electric current.

But Monday evening, about nine o'clock, occurred a mysterious interruption, which staggered all on board. Suddenly the electrical continuity was lost.

The cable was not broken, but it ceased to work. Here was a mystery. De Sauty tried it, and Professor Morse tried it. But neither could make it work. It seemed that all was over. The electricians gave it up, and the engineers were preparing to cut the cable, and to endeavor to wind it in, when suddenly *the electricity came back again.* This made the mystery greater than ever. It had been interrupted for two hours and a half. This was a phenomenon which has never been explained. Professor Morse was of opinion that the cable, in getting off the wheels, had been strained so as to open the gutta-percha, and thus destroy the insulation. If this be the true explanation, it would seem that on reaching the bottom the seam had closed, and thus the continuity had been restored. But it was certainly an untoward incident, which " cast ominous conjecture on the whole success," as it seemed to indicate that there were at the bottom of the sea causes which were wholly unknown and against which it was impossible to provide.

The return of the current was like life from the dead. Says Mullaly: "The glad news was soon circulated throughout the ship, and all felt as if they had been imbued with a new life. A rough, weather-beaten old sailor, who had assisted in coiling many a long mile of it on board the Niagara, and who was among the first to run to the telegraph office to have the news confirmed, said he would have given fifty dollars out of his pay to have saved that cable. 'I have watched nearly every mile of it,' he added, 'as it came over the side, and I would have given fifty dollars, poor a man as I am, to have saved it, although I don't expect to make any thing by it when it is laid down.' In his own simple way he expressed the feelings of every one on board, for all are as much interested in the success of the enterprise as the largest shareholder in the Company. They talked of the cable as they would of a pet child, and never was child treated with deeper solicitude than that with which the cable is watched by them. You could see the tears standing in the eyes of some as they almost cried for joy, and told their messmates that it was all right." It was indeed a great relief; and though still anxious, after watching till past midnight, a few crept to their couches, to snatch an hour or two of broken sleep. But before the morning broke, the hopes thus revived were again and finally destroyed.

It seems that the cable was running out freely at the rate of six miles an hour, while the ship was advancing but about four. This was supposed to be owing to a powerful under-current. To check this waste, the engineer applied the brakes firmly, which at once stopped the machine. The effect was to bring a heavy strain on the cable that was in the water.

The stern of the ship was down in the trough of the sea, and as it rose upward on the swell, the pressure was too great, and the cable parted.

Instantly ran through the ship a cry of grief and dismay. She was stopped in her onward path, and in a few minutes all gathered on deck with feelings which may be imagined. One who was present wrote: "The unbidden tear started to many a manly eye. The interest taken in the enterprise by all, every one, officers and men, exceeded any thing I ever saw, and there is no wonder that there should have been so much emotion at our failure." Captain Hudson says: "It made all hands of us through the day like a household or family which had lost their dearest friend, for officers and men had been deeply interested in the success of the enterprise."

There was nothing left but to return to England. The course decided upon, is thus stated in a private letter of Mr. Field, which, though intended only for his family, may be interesting to others:

"H. M. Steamer Leopard, Thursday,
August 13, 1857.

"The successful laying down of the Atlantic Telegraph Cable is put off for a short time, but its final triumph has been fully proved, by the experience that we have had since we left Valentia. My confidence was never so strong as at the present time, and I feel sure, that with God's blessing, we shall connect Europe and America with the electric cord.

"After having successfully laid—and part of the time while a heavy sea was running—three hundred and thirty-five miles of the cable, and over one hundred miles of it in water more than two miles in depth, the brakes were applied more firmly, by order of Mr. Bright, the engineer, to prevent the cable from going out so fast, and it parted.

"I retired to my state-room at a little after midnight Monday, all going on well, and at a quarter before four o'clock on Tuesday morning, the eleventh instant, I was awoke from my sleep by the cry of Stop her, back her! and in a moment Mr. Bright was in my room, with the sad intelligence that the cable was broken. In as short a time as possible I was dressed, and on deck; and Captain Hudson at once signaled the other steamers that the cable had parted, and in a few moments Captain Wainwright, of the Leopard, and Captain Sands, of the Susquehanna, were on board of the Niagara.

"I requested Captain Wainwright, the Commander of the English Telegraph Fleet, to order the Agamemnon to remain with the Niagara and Susquehanna, in this deep part of the Atlantic for a few days, to try certain experiments which will be of great value to us, and then sail with them back to England, and all wait at Plymouth until further orders. I further requested Captain Wainwright to order the Cyclops to sound here where the cable parted, and then steam back to Valentia, with letters from me to Dr. Whitehouse, and Mr. Saward, the Secretary of the Atlantic Telegraph Company; and that he should take me in the Leopard as soon as possible to Portsmouth.

" All of my requests were cheerfully complied with, and in a few hours the Cyclops had sounded, and found the bottom at two thousand fathoms, and was on her way back to Valentia with letters from me; the Niagara and the Agamemnon were connected together by the cable, and engaged in trying experiments; the Susquehanna in attendance, and the Leopard, with your affectionate ——— on board, on her way back to England.

" In my letter to Dr. Whitehouse I requested him to telegraph to London, and have a special meeting of the Directors called for twelve o'clock on Saturday, to decide whether we would have more cable made at once, and try again this season, or wait until next year.

" I shall close this letter on board, so as to have it ready to mail the moment we arrive at Portsmouth, as I wish to leave by the very next train for London, so as to be there in time to meet the Directors Saturday noon, and read them, my report, which I am busy making up.

"Do not think that I feel discouraged, or am in low spirits, for I am not; and I think I can see how this accident will be of great advantage to the Atlantic Telegraph Company.

"All the officers and men on board of the Telegraph Fleet, seem to take the greatest interest in our enterprise, and are very desirous to go out in the ships the next time.

"Since my arrival, I have received the greatest kindness and attention from all whom I have met, from the Lord Lieutenant of Ireland, down to the cabin-boys and sailors. The inclosed letter from the Knight of Kerry, I received with a basket of hothouse fruit, just as we were getting ready to leave Valentia harbor.   Your ———

"CYRUS W. FIELD."

The day that this was written, Mr. Field landed at Portsmouth, and at once hastened to London to meet the Directors. At first it was a question if they should renew the expedition this year. But their brief experience had shown the need of more ample preparations for their next attempt. They required six hundred miles more of cable to make up for over three hundred lost in the sea, and to provide a surplus so as to run no risk of falling short from other accidents; and especially they needed better machinery to pay out the cable into the ocean. These preparations required time, and before they could be made, it would be late in the autumn. Hence they reluctantly decided to defer the expedition till another year. The Niagara and the Agamemnon therefore discharged their cable at Plymouth, whence the Niagara returned home; and Mr. Field, after remaining a few weeks in London to complete the preparations for the next year, sailed for America.

He returned to find that a commercial hurricane had swept over the country, in which a thousand stately fortunes had gone down, and in which the wealth he had accumulated by years of toil had nearly suffered shipwreck. Such were the tidings that met him on landing. It had been a year of disappointments in England and America—of disasters on land and sea—and all his high hopes were

In the deep bosom of the ocean buried.

CHAPTER IX.

PREPARATIONS FOR A SECOND EXPEDITION. MR. FIELD IS MADE THE GENERAL MANAGER OF THE COMPANY. MR. EVERETT AND THE PAYING-OUT MACHINE. THE VALORUOS TAKES THE PLACE OF THE SUSQUEHANNA. THE SQUADRON ASSEMBLE AT PLYMOUTH. THEY GO TO SEA, JUNE 10. HEAVY GALE. THE AGAMEMNON IN DANGER OF BEING FOUNDERED. THE CABLE LOST THREE TIMES. THE SHIPS RETURN TO ENGLAND. MEETING OF THE DIRECTORS. SHALL THEY ABANDON THE PROJECT? ONE MORE TRIAL.

ALTHOUGH the Expedition of 1857 really advanced the project of an Atlantic Telegraph—since it was an experiment on a grand scale, and taught many lessons which could only be learned by experiment—still the effect was to weaken public confidence. Hitherto the enterprise had been accepted by the people of England and of America—almost without considering its magnitude and difficulty. They had taken it for granted as a thing which must some day be accomplished by human skill and perseverance. This confidence led them to embark their means in it. But now it had been tried and failed. This first expedition opened their eyes to the vastness of the undertaking, and led many to doubt who did not doubt before. This decline of popular faith was felt as soon as they began to call for more money. People reasoned that if the former attempt was but an experiment, it was rather a costly one. The loss of three hundred and thirty-five miles of cable, with the postponement of the expedition to another year, was equivalent to a loss of a hundred thousand pounds. To make this good, the Directors had to enlarge the capital of the Company. This new capital was not so readily obtained. Those who had subscribed before, thought they had lost enough; and the public stood aloof till they could see the result of the next "experiment." The projectors found that it was easy to go with the current of popular enthusiasm, but very hard to stem a growing popular distrust. They found how great an element of success in all public enterprises is public confidence.

But against this very revulsion of feeling they had been already warned. The Earl of Carlisle the year before had cautioned them against being too sanguine of immediate results, and reminded them that "preliminary failure was even the law and condition of ultimate success." There were many who now remembered his words, and on whom the lesson was not lost.

But whatever the depression at the failure of the first attempt to lay a telegraph across the ocean, and at the thick-coming disasters on land and sea, it did not interfere with renewed and vigorous efforts to prepare for a second expedition. The Directors gave orders for the manufacture of seven hundred miles of new cable, to make up for the loss of the previous year, and to provide a surplus against all contingencies. And the Government promised again its powerful aid.

In America, Mr. Field went on to Washington to ask a second time the use of the ships, which had already represented the country so well. He made also a special request for the services of Mr. William E. Everett. This gentleman had been the Chief-Engineer of the Niagara the year before. He had watched closely the paying-out machine, as it was put together on the deck, and with the eye of a practised mechanic, he saw that it would never do. It was too cumbrous, had too many wheels, and especially its brakes shut down with a gripe that would snap the strongest chain cable. Mr. Field saw that this was the man to remedy the defects of the old machine, and to make one that would work more smoothly. He therefore applied especially for his services. To the credit of the administration, it granted both requests in the most handsome manner. "There," said the Secretary of the Navy, handing Mr. Field the official letter, " I have given you all you asked."

After such an answer he did not wait long. The letter is dated the thirtieth of December, and in just one week, on the sixth of January, he sailed in the Persia for England with Mr. Everett. Scarcely had he arrived in London before he was made the General Manager of the Company, with control of the entire staff, including electricians and engineers. The following extract from the minutes of the Board of Directors, dated January 27, 1858, explains the new position to which he was invited:

"The Directors having for several months felt that it would greatly advance the interests of this enterprise, if Mr. Cyrus W. Field, of New-York, could be induced to come over to England, for the purpose of undertaking the general management and supervision of all the various arrangements that would be required to be carried out before the sailing of the next expedition; application was made to Mr. Field, with the view of securing his consent to the proposal, and he arrived in this country on the sixteenth instant, when it was ascertained that he would be willing, if unanimously desired by the Directors, to act in behalf of the Company as proposed; and Mr. Field having retired, it was unanimously resolved to tender him, in respect to such services, the sum of £1000 over and above his travelling and other expenses, as remuneration."

This resolution was at once communicated to Mr.

Field, who replied that he would undertake the duties of General Manager, but declined the offer of £1000, preferring to give his services to the Company without any compensation. Whereupon the Directors immediately passed another resolution:

"That Mr. Field's kind and generous offer be accepted by this Board; and that their best thanks are hereby tendered to him for his devotion to the interests of this undertaking."

The following, passed a few weeks later, on the twenty-sixth of March, was designed to emphasize the authority given over all the employés of the Company:

"*Resolved*, That Mr. Cyrus W. Field, General Manager of the Company, is hereby authorized and empowered to give such directions and orders to the officers composing the staff of the Company, as he may from time to time deem necessary and expedient with regard to all matters connected with the business proceedings of the Company, subject to the control of the Directors.

"*Resolved*, That the staff of the Company be notified hereof, and required to observe and follow such directions as may be issued by the General Manager."

As Mr. Field was thus invested with the entire charge of the preparations for the next expedition, he was made responsible for it, and felt it due alike to himself and to the Company to omit no means to insure success. It was therefore his duty to examine into every detail. The manufacture of the new cable was already under way, and there was no opportunity to make any change in its construction, even if any had been desired. But there was another matter which was quite as important to success—the construction of the paying-out machines. This had been the great defect of the previous year, and, while it continued, would render success almost impossible. No matter how many hundreds or thousands of miles of cable might be made, if the machinery was not fitted to pay it out into the sea, it would be constantly broken. To remedy these defects was an object of anxious solicitude, and to this the new manager gave his first attention. Hardly was he in London before Mr. Everett was installed at the large machine works of Easton and Amos, in Southwark, where, surrounded by plans and models, he devoted himself for three months to studying out a better invention for this most important work. At the end of that time he had a model complete, and invited a number of the most eminent engineers of London to witness its operation. Among these were Mr. Brunel, and Messrs. Lloyd, Penn, and Field, who had given the enterprise the benefit of their counsel for months, refusing all compensation; Mr. Charles T. Bright, the engineer of the Company, and his two assistants, Mr. Canning and Mr.

Clifford, and Mr. Follansbee, Chief-Engineer of the Niagara, the place which Mr. Everett had occupied the year before. The machine was set in motion, and all saw its operation, while Mr. Everett explained its parts, and the difficulties which he had tried to overcome. It was obvious at a glance that it was an immense improvement on that of the former year. It was much smaller and lighter. It would take up only about one third as much room on the deck, and had only one fourth the weight of the old machine. Its construction was much more simple. Instead of four heavy wheels, it had but two, and these were made to revolve with ease, and without danger of sudden check, by the application of what were known as self-releasing brakes. These were the invention of Mr. Appold, of London, a gentleman of fortune, but with a strong taste for mechanics, which led him to spend his time and wealth in exercising his mechanical ingenuity. These brakes were so adjusted as to bear only a certain strain, when they released themselves. This ingenious contrivance was applied by Mr. Everett to the paying-out machinery. The strength of the cable was such that it would not break except under a pressure of a little over three tons. The machinery was so adjusted that not more than half that strain could possibly come upon the cable, when the brakes would relax their grasp, the wheels revolve easily, and the cable run out into the sea "at its own sweet will." The paying-out machine, therefore, we are far from claiming as wholly an American invention. This part of the mechanism was English. The merit of Mr. Everett lay in the skill with which he adapted it to the laying of the Atlantic cable, and in his great improvements of other parts of the machinery. The whole construction, as it afterwards stood upon the decks of the Niagara and the Agamemnon, was the combined product of English and American invention. The engineers, who now saw it for the first time, were delighted. It seemed to have the intelligence of a human being, to know when to hold on, and when to let go. All felt that the great difficulty in laying the cable was removed, and that under this gentle manipulation it would glide easily and smoothly from the ship into the sea.

While these preparations were going on in London, the Niagara, which did not leave New-York till the ninth of March, arrived at Plymouth, under command of Captain Hudson, to take on board her share of the cable. Both ships had discharged their burden at Keyham Docks, where the precious freight was passed through a composition of tar and pitch and linseed-oil and beeswax, to preserve it from injury, and then had been coiled under cover to be kept safely through the winter. The Agamemnon was already at Plymouth, having been designated by the Admiralty again to take part in the work—though under a new commander, Captain George W. Preedy, a very excellent officer. The place of the Leopard was taken by the Gorgon, under command of Captain Dayman, who had made the deep-sea soundings in the Cyclops the year before.

While the English Government was thus prompt in furnishing its ships, news arrived from America that the Company could not have again the assistance of the Susquehanna, which had accompanied the Niagara the year before. She was in the West-Indies, and the yellow fever had broken out on board. What should be done? It was late to apply again to the American Government, and it was doubtful what would be the result of the application. This threatened some embarrassment. Mr. Field resolved the difficulty in a way which showed his confidence in the great and generous Government on the other side of the water, with which he had occasion so often to deal. Without waiting for the action of the Company, he called a cab, and drove straight to the Admiralty, and sent in his card to Sir John Pakington, then first Lord of the Admiralty. This gentleman, like his predecessor, Sir Charles Wood, had shown the most friendly interest in the Atlantic Telegraph, and given it his warmest support. Mr. Field was received at once, and began with true American eagerness: "I am ashamed to come to you, after what the English Government has done for the Atlantic Telegraph. But here is our case. We are disappointed in the Susquehanna. She is in the West-Indies, with the yellow fever on board. She cannot come to England to take part in the expedition. Can you do any thing for us?" Sir John replied that the Government had not ships enough for its own use; that it was at that very moment chartering vessels to take troops to Malta—"but he would see what he could do." In an hour or two he sent word to the office of the Company, that Her Majesty's ship Valorous—commanded by Captain W. C. Aldham, an officer of great experience—had been ordered to take the place of the Susquehanna in the next expedition. We mention this little incident, not so much to illustrate Mr. Field's prompt and quick manner of deciding and acting, as to show the noble and generous spirit in which the English Government responded to every appeal.

The reshipping of the cable at Plymouth occupied the whole month of April and part of May. Some changes were made in the mode adopted, it being coiled around large cones. The work was done as before, by a hundred and sixty men detailed for the purpose, of whom one fourth were the workmen of the Company, and the rest sailors who had volunteered for the duty. These were divided into gangs of forty, that relieved each other, by which the work went on day and night. In this way they coiled about thirty miles in the twenty-four hours. Owing to the increased length of cable, and the greater care in coiling, it took a longer time than the year before. The whole was completed about the middle of May. There was then in all, on board the two ships, a little over three thousand miles. This included, besides seven hundred miles of new cable, thirty-nine miles of that lost the year before, which had been recovered by the company, and a few miles of condemned cable from Greenwich, which was put on board for experiments. The shipment being thus complete, and the paying-out machines in position, the ships were ready to make a trial trip, preparatory to their final departure.

For this purpose the telegraphic squadron sailed from Plymouth on Saturday, the twenty-ninth of May, and bore southward two or three hundred miles, till the green color of the sea changing to a deep blue, showed that they had reached the great depths of the ocean. They were now in the waters of the Bay of Biscay, where the soundings were over twenty-five hundred fathoms. Here the Niagara and the Agamemnon were hung together by a hawser, being about a quarter of a mile apart. Then the cable was passed from one to the other, and a series of experiments began, designed to test both the strength of the cable and the working of the machinery. Two miles of the cable were paid out, when it parted. This would have seemed a bad sign, had it been any other part of the cable than that which was known to be imperfect and had long since been condemned. The next day three miles were paid out. This, too, was broken, but only when they tried to haul it in, and under a pressure of several tons.

Other experiments were tried, such as splicing the cable, and lowering it to the bottom of the sea—an operation which it was thought might be critical in mid-ocean, but which was performed without difficulty —and running out the cable at a rapid rate, when the speed of the ship was increased to seven knots, without causing the cable to break, or even to kink. On the whole, the result of the trip was regarded as satisfactory. The paying-out machine of Mr. Everett worked well, and the electric continuity through the whole cable was perfect. After these experiments the squadron returned to Plymouth.

Two days after, Saturday, the fifth of June, it was the fortune of the writer to arrive in Plymouth, and to witness the final preparations for the departure of the expedition. It was his privilege to attend divine service on board the Niagara the last Sabbath before she sailed. Never can he forget that day,

> So pure, so calm, so bright,
> The bridal of the earth and sky.

The squadron lay all together in Plymouth Sound. Looking at those great ships, turned from their office of war to be messengers of peace and good-will, they seemed truly "beautiful upon the waters as are the feet upon the mountains of those who publish the gospel of peace."*

Among the matters of *personal* solicitude and anxiety at this time—next to the success of the expedition—was Mr. Field himself. He was working with an activity which was unnatural—which could only be kept up by great excitement, and which involved the most serious danger. The strain on the man was more than the strain on the cable, and we were in fear that both would break together. Often he had no sleep, except such as he caught flying on the railway. Indeed, when we remonstrated, he said he could rest better there than anywhere else, for then he was not tormented with the thought of any thing undone. For the time being he could do no more; and then putting his head in the cushioned corner of the carriage, he got an hour or two of broken sleep.

Of this activity we had an instance while in Plymouth. The ships were then lying in the Sound, only waiting orders from the Admiralty to go to sea; but some business required one of the Directors to go to Paris, and as usual, it fell upon him. He left on Sun-

* Speech of the Earl of Carlisle at Valentia the year before.

day night and went to Bristol, and thence, by the first morning train, to London. Monday he was busy all day, and that night went to Paris. Tuesday, another busy day, and that night back to London. Wednesday, occupied every minute till the departure of the Great Western train. That night back to Plymouth. Thursday morning on board the Niagara, and immediately the squadron sailed.

It was the tenth day of June that the expedition left England, with fair skies and bright prospects. In truth, it was a gallant sight, as these four ships stood out to sea together—those old companions, the Niagara and the Agamemnon, leading the way, followed by their new attendants, the Valorous and the Gorgon. Never did a voyage begin with better omens. The day was one of the mildest of June, and the sea so still, that one could scarcely perceive, by the motion of the ship, when they passed beyond the breakwater off Plymouth harbor into the channel, or into the open sea. At night, it was almost a dead calm. The second day was like the first. There was scarcely wind enough to swell the sails. The ships were all in sight, and as they kept under easy steam, they seemed bound on a voyage of pleasure, gliding over a summer sea to certain success.

It had been supposed that the expedition of this year would have a great advantage over the last, from sailing two months earlier, at what was considered a more favorable season. So said all the wise men of the sea. They had given their opinion that June was the best month for crossing the Atlantic. Then they were almost sure of fair weather. The first three days of the voyage confirmed these predictions, and they who had made them, being found true prophets, shook their heads with great satisfaction.

But alas! for the vanity of human expectations, or for those who put trust in the "treacherous sea." On Sunday it began to blow. The barometer fell, and all signs indicated to the eye of a seaman rough weather. From this time they had a succession of gales for more than a week. From day to day it blew fiercer than before, till Sunday, the twentieth, when the gale was at its height, and the spirit of the storm was out on the Atlantic. Up to this time the Niagara and the Agamemnon (though they had long since parted from the Valorous and the Gorgon) had managed to keep in sight of each other; and now from the deck of the former the latter was seen a mile and a half distant, rolling heavily in the sea. The signals which she made showed that she was struggling with the fury of the gale. She was really in great danger of foundering. But this was owing, not merely to the severity of the gale, but to the enormous weight she carried, and to the way this huge bulk was stowed in the ship.

Only a few days before this we had been on board of her, and Captain Preedy showed us, in one coil, thirteen hundred miles of cable! This made a dead weight of as many hundred tons, which rendered her in rough weather almost unmanageable. To make the matter worse, she had another coil of about two hundred and fifty tons on the forward deck, where it made the head of the ship heavy. In her tremendous rolls, this coil broke loose, and threatened at a time to dash like an avalanche through the side of the ship. But at the most fearful moments the gallant seaman in command never lost his presence of mind. He was always on deck, watching with a vigilant eye the raging of the tempest, and issuing his orders with coolness and prompt decision. To this admirable skill was due the safety of the ship, and of all on board.*

But all things have an end; and this long gale at last blew itself out, and the weary ocean rocked itself to rest. Toward the last of the week the squadron

---

* As there is no trouble without a compensation, it is something that this voyage, fearful as it was, furnished a subject for a description of marvellous power. The letter to the London Times, written by Mr. Woods, its correspondent on board the Agamemnon, is one of the finest descriptions of a storm at sea we know of in the language. It is a wonderful specimen of "word-painting," and in the vividness with which it brings the scene before us, may be compared with the marine paintings of Stanfield or Turner.

got together at the appointed rendezvous in mid-ocean. As the ships came in sight, the angry sea went down; and on Friday, June twenty-fifth, just fifteen days from Plymouth, they were all together, as tranquil in the middle of the Atlantic as if in Plymouth Sound. "This evening the four vessels lay together side by side, and there was such a stillness in the sea and air, as would have seemed remarkable in an inland lake; on the Atlantic, and after what we had all so lately witnessed, it seemed almost unnatural." The boats were out, and the officers were visiting from ship to ship, telling their experiences of the voyage, and forming their plans for the morrow. Captains Aldham and Dayman said it was the worst weather they had ever experienced in the North-Atlantic. But it was the Agamemnon that suffered most. The rough sea had shaken not only the ship, but the cable in her. The upper part of the main coil had shifted, and become so twisted and tangled, that a hundred miles had to be got out and coiled in another part of the ship, so that it was not till the afternoon of Saturday, the twenty-sixth, that the splice was finally made, and the cable lowered to the bottom of the sea. The ships were then got under weigh, but had not gone three miles, before the cable broke, being caught in the machinery on board the Niagara. It was fortunate they had gone no further. Both ships at once turned about and spliced again the same afternoon, and made a fresh start. Now all went well. The paying-out machines worked smoothly, and the cable ran off easily into the sea. Thus each ship had paid out about forty miles when suddenly the current ceased!

Says the writer on the Agamemnon: "At half-past three o'clock [Sunday morning] forty miles had gone, and nothing could be more perfect and regular than the working of every thing, when suddenly Professor Thompson came on deck, and reported a total break of continuity; that the cable in fact had parted, and, as was believed at the time, from the Niagara. In another instant a gun and a blue-light warned the Valorous of what had happened, and roused all on board the Agamemnon to a knowledge that the machinery was silent, and that the first part of the Atlantic Cable had been laid and lost effectually."

This was disheartening, but not so much from the fact of a fresh breaking of the cable, as from the mystery as to its cause. The fact, of course, was known instantly on both ships, but the cause was unknown. Those on each ship supposed it had occurred on the other. With this impression, they turned about to beat up again toward the rendezvous. It was noon of Monday, the twenty-eighth, before the Agamemnon rejoined the Niagara; and then, says the writer already quoted: "While all were waiting with impatience for her explanation of how they broke the cable, she electrified every one by running up the interrogatory: 'How did the cable part?' This was astounding. As soon as the boats could be lowered, Mr. Cyrus Field, with the electricians from the Niagara, came on board, and a comparison of logs showed the painful and mysterious fact that, *at the same second of time*, each vessel discovered that a total fracture had taken place at a distance of certainly not less than ten miles from each ship; in fact, as well as can be judged, at the bottom of the ocean. That of all the many mishaps connected with the Atlantic Telegraph, this is the worst and most disheartening is certain, since it proves that, after all that human skill and science can effect to lay the wire down with safety has been accomplished, there may be some fatal obstacles to success at the bottom of the ocean, which can never be guarded against; for even the nature of the peril must always remain as secret and unknown as the depths in which it is to be encountered."

But it was no time for useless regrets. Once more the cable was joined in mid-ocean, and dropped to its silent bed, and the Niagara and the Agamemnon began to steam away toward opposite shores of the Atlantic. This time the experiment succeeded better than before. The progress of the English ship is thus reported:

"At first, the ship's speed was only two knots, the cable going three and three and a half, with a strain of fifteen hundred pounds. By and by, however, the speed was increased to four knots, the cable going five, at a strain of two thousand pounds. At this rate it was kept, with trifling variations, throughout almost the whole of Monday night, and neither Mr. Bright, Mr. Canning, nor Mr. Clifford ever quitted the machines for an instant. Toward the middle of the night, while the rate of the ship continued the same, the speed at which the cable paid out slackened nearly a knot an hour, while the dynamometer indicated as low as thirteen hundred pounds. This change could only be accounted for on the supposition that the water had shallowed to a considerable extent, and that the vessel was, in fact, passing over some submarine Ben Nevis or Skiddaw. After an interval of about an hour, the strain and rate of progress of the cable again increased, while the increase of the vertical angle seemed to indicate that the wire was sinking down the side of a declivity. Beyond this, there was no variation throughout Monday night, or, indeed, through Tuesday."

On board the Niagara was the same scene of anxious watching every hour of the day and night. Engineers and electricians were constantly on duty. "The scene at night was beautiful. Scarcely a word was her explanation of how they broke the cable, she electrified every one by running up the interrogatory: 'How did the cable part?' This was astounding. As soon as the boats could be lowered, Mr. Cyrus Field, with the electricians from the Niagara, came on board, and a comparison of logs showed the painful and mysterious fact that, *at the same second of time*, each vessel discovered that a total fracture had taken place at a distance of certainly not less than ten miles from each ship; in fact, as well as can be judged, at the bottom of the ocean. That of all the many mishaps connected with the Atlantic Telegraph, this is the worst and most disheartening is certain, since it proves that, after all that human skill and science can effect to lay the wire down with safety has been accomplished, there may be some fatal obstacles to success at the bottom of the ocean, which can never be guarded against; for even the nature of the peril must always remain as secret and unknown as the depths in which it is to be encountered."

But it was no time for useless regrets. Once more the cable was joined in mid-ocean, and dropped to its silent bed, and the Niagara and the Agamemnon began to steam away toward opposite shores of the Atlantic. This time the experiment succeeded better than before. The progress of the English ship is thus reported:

"At first, the ship's speed was only two knots, the cable going three and three and a half, with a strain of fifteen hundred pounds. By and by, however, the speed was increased to four knots, the cable going five, at a strain of two thousand pounds. At this rate it was kept, with trifling variations, throughout almost the whole of Monday night, and neither Mr. Bright, Mr. Canning, nor Mr. Clifford ever quitted the machines for an instant. Toward the middle of the night, while the rate of the ship continued the same, the speed at which the cable paid out slackened nearly a knot an hour, while the dynamometer indicated as low as thirteen hundred pounds. This change could only be accounted for on the supposition that the water had shallowed to a considerable extent, and that the vessel was, in fact, passing over some submarine Ben Nevis or Skiddaw. After an interval of about an hour, the strain and rate of progress of the cable again increased, while the increase of the vertical angle seemed to indicate that the wire was sinking down the side of a declivity. Beyond this, there was no variation throughout Monday night, or, indeed, through Tuesday."

On board the Niagara was the same scene of anxious watching every hour of the day and night. Engineers and electricians were constantly on duty. "The scene at night was beautiful. Scarcely a word was spoken; silence was commanded, and no conversation allowed. Nothing was heard but the strange rattling of the machine as the cable was running out. The lights about deck and in the quarter-deck circle added to the singularity of the spectacle; and those who were on board the ship describe the state of anxious suspense in which all were held as exceedingly impressive."

Warned by repeated failures, they hardly dared to hope for success in this last experiment. And yet the spirits of all rose, as the distance widened between the ships. A hundred miles were laid safely—a hundred and fifty—two hundred. Why might they not lay two thousand? So reasoned the sanguine and hopeful when, Tuesday night, came the fatal announcement that the electric current had ceased to flow. It afterward appeared that the cable had broken about twenty feet from the stern of the Agamemnon.

As the cable was now useless, it only remained to cut it from the stern of the Niagara. Before doing this, Mr. Field thought it a good opportunity to test its strength. For this purpose the brakes were shut down, so that the paying-out machine could not move. But still the cable did not break, although the whole weight of the Niagara hung upon that slender cord, and though several men got upon the brakes. Says Captain Hudson: "Although the wind was quite fresh, *the cable held the ship for one hour and forty minutes before breaking, and notwithstanding a strain of four tons."*

Though not unexpected, this last breaking of the cable was a sad blow to all on board. It was the end of their hopes, at least for the present expedition. Before separating, it had been agreed, that if the cable should part again before either ship had run a hundred miles, they should return and renew the attempt. If they had passed that limit, they were all to sail for Ireland. But the Niagara had run out a hundred and eleven miles, and knowing that the Agamemnon had done about the same, she expected the latter would keep on her course eastward, not stopping till she reached Queenstown. The Niagara, therefore, reluctantly bore away for the same port.

Of course, the return voyage was "any thing but gay." When soldiers come home from the war, they march with a proud step, if they have had a victorious campaign. But it is otherwise when they come with a sad tale of disaster and defeat. Seldom had an expedition begun with higher hopes, or ended in more complete failure. Who could help feeling keenly this fresh disappointment? Even with all the courage "that may become a man," heightened by a natural buoyancy of spirits, how was it possible to resist the impression of the facts they had just witnessed? If—as Lord Carlisle had told them the year before—" there was almost enough of glory in the very design of an Atlantic telegraph," that glory might still be theirs. But apparently they could hope for nothing more. They had done all that men could do. But Fate seemed against them; and who can fight against destiny? No one can blame them if they sometimes had sore misgivings, and looked out sadly upon the sea that had baffled their utmost skill, and now laughed their efforts to scorn.

In this mood they entered once more the harbor of Queenstown. The Niagara was the first to arrive and to bring tidings of the great disaster. The Agamemnon came in a few days after. Knowing the fatal impression their report was likely to produce, Mr. Field hastened to London to meet the Directors. It was high time. The news had reached there before him, and had already produced its effect. Under its impression the Board was called together. It met in the same room where, six weeks before, it had discussed the prospects of the expedition with full confidence of success. Now it met, as a council of war is summoned after a terrible defeat, to decide whether to surrender or to try once more the chances of battle. When the Directors came together, the feeling—to call it by the mildest name— was one of extreme discouragement. They looked blankly in each other's faces. With some, the feeling was one almost of despair. Sir William Brown, of

Liverpool, the first Chairman, wrote, advising them to sell the cable. Mr. Brooking, the Vice-Chairman, who had given more time to it than any other Director, when he saw that his colleagues were disposed to make still another trial, left the room, and the next day sent in his resignation, determined to take no further part in an undertaking which had been proved hopeless, and to persist in which seemed mere rashness and folly.

But others thought there was still a chance. Like Robert Bruce, who, after twelve battles and twelve defeats, yet believed that a thirteenth *might* bring victory, they clung to this bare possibility. Mr. Field and Professor Thomson gave the results of their experience, from which it appeared that there was no obstacle in the nature of the case which might not be overcome. To be sure, it was a forlorn hope. But the ships were there. Though they had lost three hundred miles of cable, they had still enough on board to cross the sea. These arguments prevailed, and it was voted to make one more trial before the project was finally abandoned. Even though the chances were a hundred to one against them, that one might secure them final success. And so it proved. But was it their own wisdom or courage that got them the victory, or were they led by that Being whose way is in the sea, and whose path is in the great waters?

CHAPTER X.

THE SHIPS SAIL ON A SECOND EXPEDITION. THEY MEET IN MID-OCEAN. SUCCESSFUL VOYAGE OF THE NIAGARA, AND OF THE AGAMEMNON. CABLE LANDED IN IRELAND AND IN NEWFOUNDLAND

A BOLD decision needs to be followed by prompt action, lest the spirit that inspires the daring attempt be weakened by delay. When once it had been fixed that there was to be another attempt to lay the Atlantic cable, no time was lost in carrying the resolve into execution. The telegraphic fleet was lying at Queenstown. The Niagara had arrived on the fifth of July, but the Agamemnon, which, through some misunderstanding, had returned to the rendezvous in mid ocean, thus crossing the Niagara on her track, did not get in till a week later. However, all were now there, safe and sound, when the order came down from London to get ready immediately to go to sea. Not an hour was lost. The ships had barely time to take in coal and other supplies for the voyage. Mr. Field hastened from England, and Prof. Thomson from his home in Scotland, and in five days the squadron was under weigh, bound once more for the middle of the Atlantic.

It was Saturday, the seventeenth of July, that the ships left on their second expedition. As they sailed out of the Cove of Cork, it was with none of the enthusiasm which attended their departure from Valentia the year before, or even from Plymouth on the tenth of June. Nobody cheered; nobody bade them God-speed. "As the ships left the harbor, there was apparently no notice taken of their departure by those on shore, or in the vessels anchored around them; every one seemed impressed with the conviction that we were engaged in a hopeless enterprise, and the squadron seemed rather to have slunk away on some discreditable mission, than to have sailed for the accomplishment of a grand national scheme." Many even of those on board felt that they were going on a fool's errand; that the Company was possessed by a kind of insanity, of which, however, they would soon be cured by another bitter experience.

On leaving this second time, it was agreed that the squadron should not try to keep together, but each ship make its way to the given latitude and longitude which was the appointed rendezvous in mid-ocean. The Niagara, being the largest, and able to carry the most coal, kept under steam the whole way, and arrived first, and waited several days for the other ships to appear. The Valorous came next, and then the Gorgon, and, last of all, the Agamemnon, which had been saving her coal for the return voyage, and had been delayed for want of a little of that wind which, in the former expedition, she had in too great abundance. Says the English correspondent on board:

"For several days in succession there was an uninterrupted calm. The moon was just at the full, and for several nights it shone with a brilliancy which turned the sea into one silvery sheet, which brought out the dark hull and white sails of the ship in strong contrast to the sea and sky, as the vessel lay all but motionless on the water, the very impersonation of solitude and repose. Indeed, until the rendezvous was gained, we had such a succession of beautiful sunrises, gorgeous sunsets, and tranquil moonlight nights, as would have excited the most enthusiastic admiration of any one but persons situated as we were. But by us such scenes were regarded only as the annoying indications of the calm, which delayed our progress and wasted our coal. By dint, however, of a judicious expenditure of fuel, and a liberal use of the cheaper motive power of sail, the rendezvous was reached on Wednesday, the twenty-eighth of July, just eleven days after our departure from Queenstown. The rest of the squadron came in sight at nightfall, but at such a distance that it was past ten o'clock on the morning of Thursday, the twenty-ninth, before the Agamemnon joined them."

" The day was beautifully calm, so no time was to be lost before making the splice; boats were soon lowered from the attendant ships, the two vessels made fast by a hawser, and the Niagara's end of the cable conveyed on board the Agamemnon. About half-past twelve o'clock the splice was effectually made. In hoisting it out from the side of the ship the leaden sinker broke short off and fell overboard; and there being no more convenient weight at hand, a thirty-two pound shot was fastened to the splice instead, and the whole apparatus was quickly dropped into the sea without any formality, and indeed almost without a spectator, for those on board the ship had witnessed so many beginnings to the telegraphic line, that it was evident they despaired of there ever being an end to it. The stipulated two hundred and ten fathoms having been paid out, the signal to start was hoisted, the hawser cast loose, and the Niagara and Agamemnon started for the last time for their opposite destinations."

At this moment the ships were nearly in mid-ocean, but not exactly. Mr. Field, who never indulges in poetical descriptions, but always gives the figures, stating the precise latitude and longitude, and from what quarter the wind blows, and how many fathoms deep the ocean is, and how many miles of cable are on board, makes the following entry in his journal:

"Thursday, July twenty-ninth, latitude fifty-two degrees nine minutes north, longitude thirty-two degrees twenty-seven minutes west. Telegraph Fleet all in sight; sea smooth; light wind from S.E. to S.S.E., cloudy. Splice made at one P.M. Signals through the whole length of the cable on board both ships perfect. Depth of water fifteen hundred fathoms; distance to the entrance of Valentia harbor eight hundred and thirteen nautical miles, and from there to the telegraph-house the shore end of the cable is laid. Distance to the entrance of Trinity Bay, Newfoundland, eight hundred and twenty-two nautical miles, and from there to the telegraph-house at the head of the bay of Bull's Arm, sixty miles, making in all eight hundred and eighty-two nautical miles. The Niagara has sixty-nine miles further to run than the Agamemnon. The Niagara and Agamemnon have each eleven hundred nautical miles of cable on board, about the same quantity as last year."

And now, as the ships are fairly apart, and will soon lose sight of each other, we will leave the Agamemnon for the present to pursue her course toward Ireland, while we follow our own Niagara to the shores of the New World. At first of course, while all hoped for success, no one dared to expect it. At least they said afterwards that "Mr. Field was the only man on board who kept up his courage through it all." But the chances seemed many to one against them; and the warnings were frequent to excite their fears. That very evening, about sunset, all again seemed lost. We quote from Mr. Field's journal: "At forty-five minutes past seven P.M., ship's time, signals from the Agamemnon ceased, and the tests applied by the electricians showed that there was a want of continuity in the cable, but the insulation was perfect. Kept on paying out from the Niagara very slowly, and constantly applying all kinds of electrical tests until ten minutes past nine, ship's time, when again commenced receiving perfect signals from the Agamemnon." At the same moment the same experience was going on on the English ship.

The next day there was a fresh cause of alarm. It was found that the Niagara had run some miles out of her course. Comparing the distance run by observation and by patent log, there was a difference of sixteen miles and a third. With such a percentage of loss, the cable would not hold out to reach Newfoundland. This was alarming, but the explanation was obvious. The mass of iron in the ship had affected the compass, so that it no longer pointed to the right quarter of the heavens. Had the Niagara been alone on the ocean, this might have caused serious trouble. But now appeared the great advantage of an attendant ship. It was at once arranged that the Gorgon should go ahead and lead the way. As she had no cable on board, her compasses were subject to no deviation. Accordingly she took her position in the advance, keeping the line along the great circle arc, which was the prescribed route. From that moment there was no variation, or but a very slight one. The two methods of computing the distance—by log and by observation—nearly coincided, and the ship varied scarcely a mile from her course till she entered Trinity Bay.

It is not necessary to follow the whole voyage, for the record is the same from day to day. It is the same sleepless watching of the cable as it runs out day and night, and the same anxious estimate of the distance that still separates them from land. Communication is kept up constantly between the ships. Mr. Field's journal contains entries like these:

"Saturday, July thirty-first. By eleven o'clock had paid out from the Niagara three hundred miles of cable; at forty-five minutes past two received signals from the Agamemnon that they had paid out from her three hundred miles of cable; at thirty-seven minutes past five finished coil on the berth-deck, and commenced paying out from the lower deck."

"Monday, August second. The Niagara getting

light, and rolling very much; it was not considered safe to carry sail to steady ship, for in case of accident it might be necessary to stop the vessel as soon as possible. Passed and signalled the Cunard steamer from Boston to Liverpool." Same day about noon, "imperfect insulation of cable detected in sending and receiving signals from the Agamemnon, which continued until forty minutes past five, when all was right again. The fault was found to be in the wardroom, about sixty miles from the lower end, which was immediately cut out, and taken out of the circuit."

"Tuesday, August third. At a quarter-past eleven, ship's time, received signals from on board the Agamemnon, that they had paid out from her seven hundred and eighty miles of cable. In the afternoon and evening passed several icebergs. At ten minutes past nine P.M., ship's time, received signal from the Agamemnon that she was in water of two hundred fathoms. At twenty minutes past ten P.M., ship's time, Niagara in water of two hundred fathoms, and informed the Agamemnon of the same.

"Wednesday, August fourth. Depth of water less than two hundred fathoms. Weather beautiful, perfectly calm. Gorgon in sight. Sixty-four miles from the telegraph-house. Received signal from Agamemnon at noon that they had paid out from her nine hundred and forty miles of cable. Passed this morning several icebergs. Made the land off entrance to Trinity Bay at eight A.M. Entered Trinity Bay at half-past twelve. At half-past two, we stopped sending signals to Agamemnon for fourteen minutes, for the purpose of making splice. At five P.M. saw Her Majesty's steamer Porcupine [which had been sent by the British Government to Newfoundland, to watch for the telegraph ships] coming to us. At half-past seven, Captain Otter, of the Porcupine, came on board of the Niagara to pilot us to the anchorage, near the telegraph-house.*

* The spot chosen as the terminus of the Atlantic cable, with the views around it—both on the water and on land—is thus described by a correspondent:

"All who have visited Trinity Bay, Newfoundland, with one consent allow it to be one of the most beautiful sheets of water they ever set eyes upon. Its color is very peculiar—an inexpressible mingling of the pure blue ocean with the deep evergreen woodlands and the serene blue sky. Its extreme length is about eighty miles, its breadth about thirty miles, opening boldly into the Atlantic on the northern side of the island. At its south-western shore it branches into the Bay of Bull's Arm, which is a quiet, safe, and beautiful harbor, about two miles in breadth, and nine or ten in length, running in a direction north-west.

"The depth of water is sufficient for the largest vessels. The tide rises seven or eight feet, and the bay terminates in a beautiful sand-beach. The shore is clothed with dark green fir-trees, which, mixed with birch and mountain-ash, present a pleasing contrast. The land rises gradually from the water all around, so as to afford one of the most agreeable town sites in the island. You ascend only about a

"Thursday, August fifth. At forty-five minutes past one A.M., Niagara anchored. Total amount of cable paid out since splice was made, ten hundred and sixteen miles, six hundred fathoms. Total amount of

quarter of a mile from the water, and there are no longer trees, but wild grass like an open prairie. Here are found at this season myriads of the upland cranberries, upon which unnumbered ptarmigan, or the northern partridge, feed.

"The raspberry, bake-apple berry, and the whortleberry are also common. Numerous little lakes may be seen in the open, elevated ground, from which flow rivulets, affording abundance of fine trout. After ascending for about a mile and a half, you are then probably three hundred or four hundred feet above the tide, and nothing can exceed the beauty of the scene when, at one view, you behold the placid waters of both Trinity and Placentia Bays—the latter sprinkled with clusters of verdant islands.

"You can now descend westward as gradually as you came up from the Telegraph landing, to the shore of Placentia Bay, where there is an excellent harbor and admirable fisheries, skirting the shore, and the accompanying road of the land Telegraph line leading from St. John's westward through the island, to Cape Ray. At this season of the year game is very abundant. Reindeer in great numbers, bears, wolves— others very numerous, the large northern hare, foxes, wild geese, ducks, etc.

"About four miles southward of the entrance of the bay of Bull's Arm, on the shore of Placentia Bay, is situated the extraordinary La Manche lead mine, the property of the Telegraph Company, already yielding a rich supply of remarkably pure galena. The place where the cable is landed is memorable in the history of the island as the naval battle-ground between the French and English in their early struggle for the exclusive occupancy of the valuable fisheries along the coast."

distance, eight hundred and eighty-two miles. Amount of cable paid out over distance run, one hundred and thirty-four miles, six hundred fathoms, being a surplus of about fifteen per cent. At two A.M., I went ashore in a small boat, and awoke persons in charge of the telegraph-house, half a mile from landing, and informed them that the Telegraph Fleet had arrived, and were ready to land the end of the cable. At forty-five minutes past two, received signal from the Agamemnon that she had paid out ten hundred and ten miles of cable. At four A.M., delivered telegraphic dispatch for the Associated Press, to be forwarded to New-York as early in the morning as the offices of the line were open.

"At a quarter-past five A.M., telegraph cable landed. At six, end of cable carried into telegraph-house, and received very strong currents of electricity through the whole cable from the other side of the Atlantic. Captain Hudson, of the Niagara, then read prayers, and made some remarks.

"At one P.M., Her Majesty's steamer Gorgon fired a royal salute of twenty-one guns."

Thus simply was the story told, that in a few hours was to send a thrill throughout the continent.

To complete the narrative of the expedition, it is necessary to include an account of the voyage of the Agamemnon. We make some extracts from the letter to the London Times, furnished by its special correspondent. We begin at the time of junction in mid-ocean, just as the ships went sailing eastward and westward:

"For the first three hours the ships proceeded very slowly, paying out a great quantity of slack, but after the expiration of this time, the speed of the Agamemnon was increased to about five knots per hour, the cable going at about six, without indicating more than a few hundred pounds of strain upon the dynamometer. Shortly after six o'clock a very large whale was seen approaching the starboard bow at a great speed, rolling and tossing the sea into foam all around, and for the first time we felt the possibility of the supposition that our second mysterious breakage of the cable might have been caused after all by one of these animals getting foul of it under water. It appeared as if it were making direct for the cable, and great was the relief of all when the ponderous living mass was seen slowly to pass astern, just grazing the cable where it entered the water, but fortunately without doing any mischief.

"All seemed to go well up to about eight o'clock; the cable paid out from the hold with an evenness and regularity which showed how carefully and perfectly it had been coiled away; and to guard against accidents which might arise in consequence of the cable having suffered injury during the storm, the indicated strain upon the dynamometer was never allowed to go beyond seventeen hundred pounds, or less than one quarter what the cable is estimated to bear, and thus far every thing looked promising of success. But, in such a hazardous work, no one knows what a few minutes may bring forth, for soon after eight, an injured portion of the cable was discovered about a mile or two from the portion paying out. Not a moment was lost by Mr. Canning, the engineer on duty, in setting men to work to cobble up the injury as well as time would permit, for the cable was going out at such a rate that the damaged portion would be paid overboard in less than twenty minutes, and former experience had shown us that to check either the speed of the ship, or the cable, would, in all probability, be attended by the most fatal results.

"Just before the lapping was finished, Professor Thomson reported that the electrical continuity of the wire had ceased, but that the insulation was still perfect; attention was naturally directed to the injured piece as the probable source of the stoppage, and not a moment was lost in cutting the cable at that point, with the intention of making a perfect splice. To the consternation of all, the electrical tests applied showed the fault to be overboard, and in all probability some fifty miles from the ship. Not a second was to be lost, for it was evident that the cut portion must be paid overboard in a few minutes, and in the mean time, the tedious and difficult operation of making a splice had to be performed. The ship was immediately stopped, and no more cable paid out than was absolutely necessary to prevent it breaking.

"As the stern of the ship was lifted by the waves, a scene of the most intense excitement followed. It seemed impossible, even by using the greatest possible speed, and paying out the least possible amount of cable, that the junction could be finished before the part was taken out of the hands of the workmen. The main hold presented an extraordinary scene; nearly all the officers of the ship and of those connected with the expedition, stood in groups about the coil, watching with intense anxiety the cable, as it slowly unwound itself nearer and nearer the joint, while the workmen, directed by Mr. Canning, under whose superintendence the cable was originally manufactured, worked at the splice as only men could work who felt that the life and death of the expedition depended upon their rapidity. But all their speed was to no purpose, as the cable was unwinding within a hundred fathoms, and, as a last and desperate resource, the cable was stopped altogether, and, for a few minutes, the ship hung on by the end. Fortunately, however, it was only for a few minutes, as the strain was continually rising above two tons, and it would not hold on much longer; when the splice was finished, the signal was made to loose the stopper, and it passed overboard safely enough.

" When the excitement consequent upon having so narrowly saved the cable had passed away, we awoke to the consciousness that the case was still as hopeless as ever, for the electrical continuity was still entirely wanting. Preparations were consequently made to pay out as little rope as possible, and to hold on for six hours, in the hopes that the fault, whatever it might be, might mend itself before cutting the cable and returning to the rendezvous to make another splice. The magnetic needles on the receiving instruments were watched closely for the returning signals; when, in a few minutes, the last hope was extinguished by their suddenly indicating dead earth, which tended to show that the cable had broken from the Niagara, or that the insulation had been completely destroyed.

" In three minutes, however, every one was agreeably surprised by the intelligence that the stoppage had disappeared, and that the signals had again appeared at their regular intervals from the Niagara. It is needless to say what a load of anxiety this news removed from the minds of every one; but the general confidence in the ultimate success of the operations was much shaken by the occurrence, for all felt that every minute a similar accident might occur. For some time the paying-out continued as usual, but toward the morning another damaged place was discovered in the cable; there was fortunately, however, time to repair it in the hold without in any way interfering with the operations beyond for a time slightly reducing the speed of the ship.

"During the morning of Friday, the thirtieth, every thing went well; the ship had been kept at the speed of about five knots, the cable paid out at about six, the average angle with the horizon at which it left the ship being about fifteen degrees, while the indicated strain upon the dynamometer seldom showed more than sixteen hundred pounds to seventeen hundred pounds. Observations made at noon showed that we had made good ninety miles from the starting-point since the previous day, with an expenditure, including the loss in lowering the splice and during the subsequent stoppages, of one hundred and thirty-five miles of the cable. During the latter portion of the day the barometer fell considerably, and toward the evening it blew almost a gale of wind from the eastward, dead ahead of course. As the breeze freshened, the speed of the engines was gradually increased, but the wind more than increased in proportion, so that, before the sun went down, the Agamemnon was going full steam against the wind, only making a speed of about four knots an hour. During the evening topmasts were lowered, and spars, yards, sails, and indeed every thing aloft that could offer resistance to the wind, was sent down on deck; but still the ship made but little way, chiefly in consequence of the heavy sea, though the enormous quantity of fuel consumed showed us that, if the wind lasted, we should be reduced to burning the masts, spars, and even the decks, to bring the ship into Valentia.

"It seemed to be our particular ill-fortune to meet with head-winds whichever way the ship's head was turned. On our journey out we had been delayed, and obliged to consume an undue proportion of coal, for want of an easterly wind, and now all our fuel was wanted because of one. However, during the next day the wind gradually went around to the south-west, which, though it raised a very heavy sea, allowed us to husband our small remaining store of fuel.

"At noon on Saturday, the thirty-first of July, observations showed us to have made good one hundred and twenty miles of distance since noon of the previous day, with a loss of about twenty-seven per cent of cable. The Niagara, as far as could be judged from the amount of cable she paid out, which, by a previous arrangement, was signalled at every ten miles, kept pace with us, within one or two miles, the whole distance across. During the afternoon of Saturday, the wind again freshened up, and before nightfall it again blew nearly a gale of wind, and a tremendous sea ran before it from the south-west, which made the Agamemnon pitch to such an extent that it was thought impossible the cable could hold on through the night; indeed, had it not been for the constant care and watchfulness exercised by Mr. Bright, and the two energetic engineers, Mr. Canning and Mr. Clifford, who acted with him, it could not have been done at all. Men were kept at the wheels of the machine to prevent their stopping as the stern of the ship rose and fell with the sea, for, had they done so, the cable must undoubtedly have parted.

"During Sunday the sea and wind increased, and before the evening it blew a smart gale. Now, indeed, were the energy and activity of all engaged in the operation tasked to the utmost. Mr. Hoar and Mr. Moore, the two engineers who had the charge of the relieving-wheels of the dynamometer, had to keep watch and watch alternately every four hours, and while on duty durst not let their attention be removed from their occupation for one moment, for on their releasing the brakes every time the stern of the ship fell into the trough of the sea entirely depended the safety of the cable, and the result shows how ably they discharged their duty. Throughout the night, there were few who had the least expectation of the cable holding on till morning, and many remained awake listening for the sound that all most dreaded to hear—namely, the gun which should announce the failure of all our hopes. But still the cable, which, in comparison with the ship from which it was paid out, and the gigantic waves among which it was delivered, was but a mere thread, continued to hold on, only leaving a silvery phosphorous line upon the stupendous seas as they rolled on toward the ship.

"With Sunday morning came no improvement in the weather; still the sky remained black and stormy to windward, and the constant violent squalls of wind and rain which prevailed during the whole day, served to keep up, if not to augment the height of the waves. But the cable had gone through so much during the night, that our confidence in its continuing to hold was much restored.

"At noon, observations showed us to have made good one hundred and thirty miles from noon of the previous day, and about three hundred and sixty from our starting-point in mid-ocean. We had passed by the deepest sounding of twenty-four hundred fathoms, and over more than half of the deep water generally, while the amount of cable still remaining in the ship was more than sufficient to carry us to the Irish coast, even supposing the continuance of the bad weather should oblige us to pay out the same amount of slack cable we had been hitherto wasting. Thus far things looked very promising for our ultimate success. But former experience showed us only too plainly that we could never suppose that some accident might not arise until the ends had been fairly landed on the opposite shores.

"During Sunday night and Monday morning the weather continued as boisterous as ever, and it was only by the most indefatigable exertions of the engineer upon duty that the wheels could be prevented from stopping altogether, as the vessel rose and fell with the sea, and once or twice they did come completely to a stand-still, in spite of all that could be done to keep them moving; but fortunately they were again set in motion before the stern of the ship was thrown up by the succeeding wave. No strain could be placed upon the cable, of course; and though the dynamometer occasionally registered seventeen hundred pounds as the ship lifted, it was oftener below one thousand, and was frequently nothing, the cable running out as fast as its own weight and the speed of the ship could draw it. But even with all these forces acting unresistedly upon it, the cable never paid itself out at a greater speed than eight knots an hour at the time the ship was going at the rate of six knots and a half. Subsequently, however, when the speed of the ship even exceeded six knots and a half, the cable never ran out so quick. The average speed maintained by the ship up to this time, and, indeed, for the whole voyage, was about five knots and a half, the cable, with occasional exceptions, running about thirty per cent faster.

"At noon on Monday, August second, had made good one hundred and twenty-seven and a half miles since noon of the previous day, and completed more than the half way to our ultimate destination.

"During the afternoon an American three-masted schooner, which afterward proved to be the Chieftain, was seen standing from the eastward toward us. No notice was taken of her at first, but when she was within about half a mile of the Agamemnon she altered her course, and bore right down across our bows. A collision, which might prove fatal to the cable, now seemed inevitable, or could only be avoided by the equally hazardous expedient of altering the Agamemnon's course. The Valorous steamed ahead, and fired a gun for her to heave to, which, as she did not appear to take much notice of, was quickly followed by another from the bows of the Agamemnon, and a second and third from the Valorous, but still the vessel held on her course; and as the only resource left to avoid a collision, the course of the Agamemnon was altered just in time to pass within a few yards of her.

It was evident that our proceedings were a source of the greatest possible astonishment to them, for all her crew crowded upon her deck and rigging. At length they evidently discovered who we were, and what we were doing, for the crew manned the rigging, and dipping the ensign several times they gave us three hearty cheers. Though the Agamemnon was obliged to acknowledge these congratulations in due form, the feelings of annoyance with which we regarded the vessel which, either by the stupidity or carelessness of those on board, was so near adding a fatal and unexpected mishap to the long chapter of accidents which had already been encountered, may easily be imagined. To those below, who of course did not see the ship approaching, the sound of the first gun came like a thunderbolt, for all took it as the signal of the breaking of the cable. The dinner-tables were deserted in a moment, and a general rush made up the hatches to the deck; but before reaching it, their fears were quickly banished by the report of the succeeding gun, which all knew well could only be caused by a ship in our way or a man overboard.

"Throughout the greater portion of Monday morning the electrical signals from the Niagara had been getting gradually weaker, until they ceased altogether for nearly three quarters of an hour. Our uneasiness, however, was in some degree lessened by the fact that the stoppage appeared to be a want of continuity, and not any defect in insulation, and there was consequently every reason to suppose that it might arise from faulty connection on board the Niagara. Accordingly Professor Thomson sent a message to the effect that the signals were too weak to be read, and, as if they had been awaiting such a signal to increase their battery power, the deflections immediately returned even stronger than they had ever been before. Toward the evening, however, they again declined in force for a short time. With the exception of these little stoppages, the electrical condition of the submerged wire seemed to be much improved. It was evident that the low temperature of the water at the immense depth improved considerably the insulating properties of the gutta-percha, while the enormous pressure to which it must have been subjected probably tended to consolidate its texture, and to fill up any air-bubbles or slight faults in manufacture which may have existed.

"The weather during Monday night moderated a little, but still there was a very heavy sea on, which endangered the wire every second minute.

"About three o'clock on Tuesday morning, all on board were startled from their beds by the loud booming of a gun. Every one, without waiting for the performance of the most particular toilet, rushed on deck to ascertain the cause of the disturbance. Contrary to all expectation, the cable was safe, but just in the gray light could be seen the Valorous rounded to in the most warlike attitude, firing gun after gun in quick succession toward a large American bark, which, quite unconscious of our proceeding, was standing right across our stern. Such loud and repeated remonstrances from a large steam frigate were not to be despised, and, evidently without knowing the why or the wherefore, she quickly threw her sails aback and remained hove to. Whether those on board her considered that we were engaged in some fillibustering expedition, or regarded our proceedings as another British outrage upon the American flag, it is impossible to say; but certain it is that, apparently in great trepidation, she remained hove to until we had lost sight of her in the distance.

"Tuesday was a much finer day than any we had experienced for nearly a week, but still there was a considerable sea running, and our dangers were far from passed; yet the hopes of our ultimate success ran high. We had accomplished nearly the whole of the deep-sea portion of the route in safety, and that, too, under the most unfavorable circumstances possible; therefore there was every reason to believe that unless some unforeseen accident should occur, we should accomplish the remainder.

"About five o'clock in the evening, the steep submarine mountain which divides the telegraphic plateau from the Irish coast was reached, and the sudden shallowing of the water had a very marked effect upon the cable, causing the strain on and the speed of it to lessen every minute. A great deal of slack was paid out to allow for any great inequalities which might exist, though undiscovered by the sounding-line. About ten o'clock the shoal water of two hundred and fifty fathoms was reached; the only remaining anxiety now was the changing from the lower main coil to that upon the upper deck, and this most difficult and dangerous operation was successfully performed between three and four o'clock on Wednesday morning.

"Wednesday was a beautiful, calm day; indeed, it was the first on which any one would have thought of making a splice since the day we started from the rendezvous. We therefore congratulated ourselves on having saved a week by commencing operations on the Thursday previous. At noon, we were eighty-nine miles distant from the telegraph station at Valentia. The water was shallow, so that there was no difficulty in paying out the wire almost without any loss of slack, and all looked upon the undertaking as virtually accomplished.

"At about one o'clock in the evening, the second change from the upper-deck coil to that upon the orlop-deck was safely effected, and shortly after the vessels exchanged signals that they were in two hundred fathoms water. As the night advanced the speed of the ship was reduced, as it was known that we were only a short distance from the land, and there would be no advantage in making it before daylight in the morning. About twelve o'clock, however, the Skelligs Light was seen in the distance, and the Valorous steamed on ahead to lead us in to the coast, firing rockets at intervals to direct us, which were answered by us from the Agamemnon, though, according to Mr. Moriarty, the master's wish, the ship, disregarding the Valorous, kept her own course, which proved to be the right one in the end.

"By daylight on the morning of Thursday, the bold and rocky mountains which entirely surround the wild and picturesque neighborhood of Valentia, rose right before us at a few miles' distance. Never, probably, was the sight of land more welcome, as it brought to a successful termination one of the greatest, but, at the same time, most difficult schemes which was ever undertaken. Had it been the dullest and most melancholy swamp on the face of the earth that lay before us, we should have found it a pleasant prospect; but, as the sun rose from the estuary of Dingle Bay, tinging with a deep, soft purple the lofty summits of the steep mountains which surround its shores, and illuminating the masses of morning vapor which hung upon them, it was a scene which might vie in beauty with any thing that could be produced by the most florid imagination of an artist.

"No one on shore was apparently conscious of our approach, so the Valorous steamed ahead to the mouth of the harbor and fired a gun. Both ships made straight for Doulus Bay, and about six o'clock came to anchor at the side of Beginish Island, opposite to Valentia. As soon as the inhabitants became aware of our approach, there was a general desertion of the place, and hundreds of boats crowded around us, their passengers in the greatest state of excitement to hear all about our voyage. The Knight of Kerry was absent in Dingle, but a messenger was immediately dispatched for him, and he soon arrived in Her Majesty's gunboat Shamrock. Soon after our arrival, a signal was received from the Niagara that they were preparing to land, having paid out one thousand and thirty nautical miles of cable, while the Agamemnon had accomplished her portion of the distance with an expenditure of one thousand and twenty miles, making the total length of the wire submerged two thousand and fifty geographical miles. Immediately after the ships cast anchor, the paddle-box boats of the Valorous were got ready, and two miles of cable coiled away in them, for the purpose of landing the end; but it was late in the afternoon before the procession of boats left the ship, under a salute of three rounds of small-arms from the detachment of marines on board the Agamemnon, under the command of Lieutenant Morris.

"The progress of the end to the shore was very slow, in consequence of the very stiff wind which blew at the time, but at about three o'clock the end was safely brought on shore at Knightstown, Valentia, by Mr. Bright and Mr. Canning, the chief and second engineers, to whose exertions the success of the undertaking is attributable, and the Knight of Kerry.* The end was immediately laid in the trench which had been dug to receive it, while a royal salute, making the neighboring rocks and mountains reverberate, announced that the communication between the Old and the New World had been completed."

* A name that occurs several times in this history, and one never to be mentioned but with honor. The Knight of Kerry is a Lord of the Isles on that part of the Irish coast; and from the constant interest which he has shown in this enterprise, and his generous hospitality to all connected with it, he has made many friends, by whom he will be remembered on both sides of the Atlantic.

CHAPTER XI.

NEWS OF THE SUCCESS. GREAT EXCITEMENT IN AMERICA. POPULAR ENTHUSIASM. CELEBRATION IN NEW-YORK AND OTHER CITIES.

WHOEVER shall write the history of popular enthusiasms, must give a large space to the Atlantic Telegraph. Never did the tidings of any great achievement—whether in peace or war—more truly electrify a nation. No doubt, the impression was the greater because it took the country by surprise. Had the attempt succeeded in June, it would have found a people prepared for it. But the failure of the first expedition, added to that of the previous year, settled the fate of the enterprise in the minds of the public. It was a very grand but hopeless undertaking; and its projectors shared the usual lot of those who conceive vast designs, and venture on great enterprises, which are not successful—to be regarded with a mixture of derision and pity.

Such was the temper of the public mind, when at noon of Thursday, the fifth of August, the following despatch was received:

"United States Frigate Niagara,
Trinity Bay, Newfoundland, August 5, 1858.

"To the Associated Press, New-York:
"The Atlantic Telegraph fleet sailed from Queenstown, Ireland, Saturday, July seventeenth, and met in mid-ocean Wednesday, July twenty-eighth. Made the splice at one P.M., Thursday, the twenty-ninth, and separated—the Agamemnon and Valorous, bound to Valentia, Ireland; the Niagara and Gorgon, for this place, where they arrived yesterday, and this morning the end of the cable will be landed.

"It is one thousand six hundred and ninety-six nautical, or one thousand nine hundred and fifty statute miles from the Telegraph House at the head of Valentia harbor to the Telegraph House at the Bay of Bulls, Trinity Bay, and for more than two thirds of this distance the water is over two miles in depth. The cable has been paid out from the Agamemnon at about the same speed as from the Niagara. The electric signals sent and received through the whole cable are perfect.

"The machinery for paying out the cable worked in the most satisfactory manner, and was not stopped for a single moment from the time the splice was made until we arrived here.

"Captain Hudson, Messrs. Everett and Woodhouse, the engineers, the electricians, the officers of the ship, and in fact, every man on board the telegraph fleet, has exerted himself to the utmost to make the expedition successful, and by the blessing of Divine Providence it has succeeded.

"After the end of the cable is landed and connected with the land line of telegraph, and the Niagara has discharged some cargo belonging to the Telegraph Company, she will go to St. John's for coal, and then proceed at once to New-York.

"CYRUS W. FIELD."

The impression of this simple announcement it is impossible to conceive. It was immediately telegraphed to all parts of the United States, and everywhere produced the greatest excitement. In some places all business was suspended; men rushed into the streets, and flocked to the offices where the news was received. An impressive scene was witnessed at a religious convocation in New-England. At Andover, Massachusetts, the news arrived while the Alumni of the Theological Seminary were celebrating their semi-centennial anniversary by a dinner. One thousand persons were present, all of whom rose to their feet, and gave vent to their excited feelings by continued and enthusiastic cheers. When quiet was restored, Rev. Dr. Adams, of New-York, said his heart was too full for a speech, and suggested, as the more fitting utterance of what all felt, that they should join in thanksgiving to Almighty God. Rev. Dr. Hawes, of Hartford, then led the assembly in fervent prayer, acknowledging the great event as from the hand of God, and as calculated to hasten the triumphs of civilization and Christianity. Then all standing up together, sang, to the tune of Old Hundred, the majestic doxology:

> "Praise God, from whom all blessings flow,
> Praise Him all creatures here below;
> Praise Him above, ye heavenly host,
> Praise Father, Son, and Holy Ghost!"

Thus—said Dr. Hawes—"we have now consecrated this new power, so far as our agency is concerned, to the building up of the truth."

In New-York the news was received at first with some incredulity. But as it was confirmed by subsequent dispatches, the city broke out into tumultuous rejoicing. Never was there such an outburst of popular feeling. In Boston a hundred guns were fired on the Common, and the bells of the city were rung for an hour to give utterance to the general joy. Similar scenes were witnessed in all parts of the United States. I have now before me the New-York papers of August, 1858, and from the memorable fifth, when the landing took place, to the end of the month, they contain hardly any thing else than popular demonstrations in honor of the Atlantic Telegraph. It was indeed a national jubilee.

It was natural that this overflow of public feeling should express itself towards one who was recognized as the author of the great work, which inspired such universal joy. Mr. Field, much to his own surprise, "awoke and found himself famous." In twenty-four hours his name was on millions of tongues. Congratulations poured in from all quarters, from mayors of cities and governors of States; from all parts of the Union and the British Provinces; from the President of the United States and the Governor-General of Canada. Mr. Buchanan telegraphed to Mr. Field, at Trinity Bay:

"My Dear Sir: I congratulate you with all my heart on the success of the great enterprise with which your name is so honorably connected. Under the blessing of Divine Providence I trust it may prove instrumental in promoting perpetual peace and friendship between the kindred nations."

The popular estimate of the achievement and its author went still farther. With the natural exaggeration common to masses of men, when carried away by a sudden enthusiasm, the Atlantic Telegraph was hailed as an immense stride in the onward progress of the race, an event in the history of the world hardly inferior to the discovery of America, or to the invention of the art of printing; and the name of its projector was coupled with those of Franklin and Columbus. He who but yesterday was regarded as a visionary, to-day was exalted as a benefactor of his country and of mankind.

This avalanche of praise was quite overwhelming. It is always embarrassing to be forced into sudden conspicuity, and to find one's self the object of general attention and applause. While feeling this embarrassment, Mr. Field could not but be gratified to witness the public joy at the success of the enterprise, and he was deeply touched and grateful for the appreciation of his own services. But probably all these public demonstrations did not go to his heart so much as private letters received from the other side of the Atlantic, from those who had shared the labors of the enterprise—old companions in arms who had borne with him the heavy burden, and now were fully entitled to a share in the honor which was the reward of their common toil.

As a sample of the congratulations which came from beyond the sea, we quote a single passage from a letter of Mr. George Saward, the Secretary of the Company in London, written immediately on receiving the news of the success of the enterprise. Under the impression of that event, he writes to Mr. Field:

"At last the great work is successful. I rejoice at

it for the sake of humanity at large. I rejoice at it for the sake of our common nationalities, and last but not least, for your personal sake. I most heartily and sincerely rejoice with you, and congratulate you, upon this happy termination to the trouble and anxiety, the continuous and persevering labor, and never-ceasing and sleepless energy, which the successful accomplishment of this vast and noble enterprise has cost you. Never was man more devoted—never did man's energy better deserve success than yours has done. May you in the bosom of your family reap those rewards of repose and affection, which will be doubly sweet from the reflection, that you return to them after having been under Providence the main and leading principal in conferring a vast and enduring benefit on mankind. If the contemplation of fame has a charm for you, you may well indulge in the reflection; for the name of Cyrus W. Field will now go onward to immortality, as long as that of the Atlantic Telegraph shall be known to mankind."

The Directors, whose faith and courage had been so severely tried, now felt double joy, for their friend and for themselves, at this glorious result of their united labors. Mr. Peabody wrote that "his reflections must be like those of Columbus, after the discovery of America." Sir Charles Wood and Sir John Pakington, who, as successive First Lords of the Admiralty had supported the enterprise with the constant aid of the British Government, wrote to Mr. Field letters of congratulation on the great work which had been carried through mainly by his energy and unconquerable will. They were above any petty national jealousy, and never imagined that it would detract aught from the just honor of England, to award full praise to the courage and enterprise of an American.

On his part, Mr. Field was equally anxious to acknowledge the invaluable aid given by others—aid, without which the efforts of no single individual could command success. On his arrival at St. John's, he was welcomed with enthusiasm by the whole population. An address was presented to him by the Executive Council of Newfoundland, in which they offer their hearty congratulations on the success of the undertaking, which they recognize as chiefly due to him. "Intimately acquainted as we have been"— these are their words—" with the energy and enterprise which have distinguished you from the commencement of the great work of telegraph connection between the Old and the New Worlds; and *feeling that under Providence this triumph of science is mainly due to your well-directed and indomitable exertions*, we desire to express to you our high appreciation of your success to the cause of the world's progress," etc.; to which Mr. Field replied, recognizing in turn the cordial support which he had always received from the Government of Newfoundland. The Chamber of Commerce of St. John's also presented an address in similar terms, to which he replied — after acknowledging their kind mention of his own labors and sacrifices :

"But it would not only be ungenerous, but unjust, that I should for a moment forget the services of those who were my co-workers in this enterprise, and without whom any labors of mine would have been unavailing. It would be difficult to enumerate the many gentlemen whose scientific acquirements, and skill and energy have been devoted to the advancement of this work, and who have so mainly produced the issue which has called forth this expression of your good wishes on my behalf. But I could not do justice to my own feelings did I fail to acknowledge how much is owing to Captain Hudson and the officers of the Niagara, whose hearts were in the work, and whose toil was unceasing; to Captain Dayman of her Majesty's ship Gorgon, for the soundings so accurately made by him last year, and for the perfect manner in which he led the Niagara over the great-circle arc while laying the cable; to Captain Otter, of the Porcupine, for the careful survey made by him in Trinity Bay, and for the admirable manner in which he piloted the Niagara at night to her anchorage; to Mr. Everett, who has for months devoted his whole time to designing and perfecting the beautiful machinery that has so successfully paid out the cable from the ships—machinery so perfect in every respect, that it was not for one moment stopped on board the Niagara until she reached her destination in Trinity Bay; to Mr. Woodhouse, who superintended the coiling of the cable, and zealously and ably coöperated with his brother engineer during the progress of paying out; to the electricians for their constant watchfulness; to the men for their almost ceaseless labor; (and I feel confident that you will have a good report from the commanders, engineers, electricians, and others on board the Agamemnon and Valorous, the Irish portion of the fleet;) to the Directors of the Atlantic Telegraph Company for the time they have devoted to the undertaking without receiving any compensation for their services, (and it must be a pleasure to many of you to know that the director, who has devoted more time than any other, was for many years a resident of this place, and well known to all of you—I allude to Mr. Brooking, of London;) to Mr. C. M. Lampson, a native of New-England, but who has for the last twenty-seven years resided in London, who appreciated the great importance of this enterprise to both countries, and gave it most valuable aid, bringing his sound judgment and great business talent to the service of the Company; to that distinguished American, Mr. George Peabody, and his most worthy partner, Mr. Morgan, who not only assisted it most liberally with their means, but to whom I could always go with confidence for advice."

Such acknowledgments, constantly repeated, showed a mind incapable of envy or jealousy; that was chiefly anxious to recognize the services of others, and that they should receive from the public, both of England and America, the honors which they had so nobly earned.

After two or three days' delay at St. John's, which the Niagara was obliged to make for coal, but which the people spent in festivity and rejoicing, she left for New-York, where she arrived on the eighteenth— two weeks from the landing of the cable in Trinity Bay. These had been weeks of great excitement, yet not unmingled with suspense and anxiety. The public, eager for news, devoured every thing that concerned the telegraph with impatience, but all was not satisfactory. Dispatches from Trinity Bay said that signals were continually passing over the cable, yet no news reached the public from the other side of the Atlantic. This was partially explained by a dispatch from Mr. Field, sent from Trinity Bay to the Associated Press as early as the seventh:

"We landed here in the woods, and until the telegraph instruments are all ready, and perfectly adjusted, no communications can pass between the two continents; but the electric currents are received freely.

"You shall have the earliest intimation when all is ready, but it may be some days before every thing is perfected. The first through message between Europe and America will be from the Queen of Great Britain to the President of the United States, and the second his reply."

But as the public grew more impatient, and friends sent anxious inquiring messages, he telegraphed again from St. John's on the eleventh :

"Before I left London, the Directors of the Atlantic Telegraph Company decided unanimously that, after the cable was laid, and the Queen's and President's messages transmitted, the line should be kept for several weeks for the sole use of Dr. Whitehouse, Professor Thomson, and other electricians, to enable them to test thoroughly their several modes of telegraphing, so that the Directors might decide which was the best and most rapid method for future use; for it was considered that after the line should be once thrown open for business, it would be very difficult to obtain it for experimental purposes, even for a short time.

"Due notice will be given when the line will be ready for business, and the tariff of prices."

Still the public were not satisfied, and many were beginning to doubt, when, on the sixteenth, it was suddenly announced that the Queen's message was received. As this was between the heads of the two countries, we give both the message and the reply:

"To the President of the United States, Washington:

"The Queen desires to congratulate the President upon the successful completion of this great international work, in which the Queen has taken the deepest interest.

"The Queen is convinced that the President will join with her in fervently hoping that the electric cable which now connects Great Britain with the United States will prove an additional link between the nations, whose friendship is founded upon their common interest and reciprocal esteem.

"The Queen has much pleasure in thus communicating with the President, and renewing to him her wishes for the prosperity of the United States."

"Washington City, August 16, 1858.
"To Her Majesty Victoria, the Queen of Great Britain:

"The President cordially reciprocates the congratulations of her Majesty the Queen, on the success of the great international enterprise accomplished by the science, skill, and indomitable energy of the two countries.

"It is a triumph more glorious, because far more useful to mankind, than was ever won by conqueror on the field of battle.

"May the Atlantic Telegraph, under the blessing of Heaven, prove to be a bond of perpetual peace and friendship between the kindred nations, and an instrument destined by Divine Providence to diffuse religion, civilization, liberty, and law throughout the world.

"In this view, will not all nations of Christendom spontaneously unite in the declaration that it shall be for ever neutral, and that its communications shall be held sacred in passing to their places of destination, even in the midst of hostilities?

"JAMES BUCHANAN."

The arrival of the Queen's message was the signal for a fresh outbreak of popular enthusiasm. The next morning, August seventeenth, the city of New-York was awakened by the thunder of artillery. A hundred guns were fired in the Park at daybreak, and the salute was repeated at noon. At this hour, flags were flying from all the public buildings, and the bells of the principal churches began to ring, reminding one of Tennyson's ode to the happy Christmas bells that were ringing out the departing year:

… HISTORY OF THE ATLANTIC TELEGRAPH. 231

> Ring out the old, ring in the new,
> Ring out the false, ring in the true.

That night the city was illuminated. Never had it seen such a brilliant spectacle. It seemed as if it were intended to light up the very heavens. Such was the blaze of light around the City Hall, that the cupola caught fire, and was consumed, and the Hall itself narrowly escaped destruction. Similar demonstrations took place in other parts of the United States. From the Atlantic to the Valley of the Mississippi, and to the Gulf of Mexico, in every city was heard the firing of guns and the ringing of bells. Nothing seemed too extravagant to give expression to the popular rejoicing.

The next morning after this illumination, the Niagara entered the harbor of New-York, and Mr. Field hastened to his home. The night before leaving the ship, he had written to a late hour to the Directors in London, giving a full report of the laying of the cable, which he closed by resigning the position which he had held for the last seven months. He wrote:

"At your unanimous request, but at a very great personal sacrifice to myself, I accepted the office of General Manager of the Atlantic Telegraph Company, for the sole purpose of doing all in my power to aid you to make the enterprise successful; and as that object has been attained, you will please accept my resignation. It will always afford me pleasure to do any thing in my power, consistent with my duties to my family and my own private affairs, to promote the interests of the Atlantic Telegraph Company."

Once more with his family, Mr. Field hoped for a brief interval of rest and quiet. But this was impossible. The great event with which his name was connected was too fresh in the public mind. He could not escape public observation. He was at once thronged with visitors, offering their congratulations, and his house surrounded with crowds eager to see and hear him. While making all allowance for popular excitement, yet none could deny that a service so great demanded some public recognition. Even in England, where the enthusiasm did not approach that in this country, still the wondrous character of the achievement was fully acknowledged. Said the London Times on the morning of the sixth of August: "Since the discovery of Columbus, nothing has been done in any degree comparable to the vast enlargement which has thus been given to the sphere of human activity." "More was done yesterday for the consolidation of our empire, than the wisdom of our statesmen, the liberality of our Legislature, or the loyalty of our colonists, could ever have effected." To mark the public benefit which had been conferred, the Chief-Engineer of the Expedition, Mr. Charles T.

Bright, was knighted, and Captains Preedy and Aldham were both made Companions of the Bath. Thus England showed her appreciation of their services.

But in this country titles and honors come not from the Government, but from the people. Popular enthusiasm exhausted itself in eulogies of the man who had linked the Old World to the New. It seems strange now to sit down in cool blood and read what was published in the papers of that day. A collection of American journals issued during that eventful month, August, 1858, would be a literary curiosity.*

---

* Such a curiosity exists, prepared by the industry of a gentleman who is one of the most careful collectors of the events of his time—thus gathering up and preserving the materials of future history—Mr. John R. Bartlett, Secretary of State of Rhode Island. This gentleman has kept files of all the papers referring to the Atlantic Telegraph, from which he has compiled a very unique volume. It is in the form of a scrap-book, but on a gigantic scale, being of a size equal to Webster's large Dictionary. It is made up entirely of newspaper cuttings, classified under different heads, and neatly arranged in double columns on nearly four hundred folio pages. The matter thus compressed would make between three and four octavo volumes of the size of Prescott's Histories, if printed in the style of those works. Every thing is included that could be gathered from the European as well as American papers, touching the claims of the inventors and projectors of the electric telegraph in general, and of the Atlantic Telegraph in particular. The historical sketches are set off by illustrations taken from the pictorial papers. Altogether it embraces more of the materials of a history of this subject than any volume with which we are

Nor was it merely in such outward demonstrations that the public enthusiasm showed itself. The feeling struck deeper, and reached all minds. While the people shouted and cannon roared, sober and thoughtful men pondered on the change that was being wrought in the earth. Business men reasoned how it would affect the commerce of the world, while the philanthropic regarded it as the forerunner of an age of universal peace. The first message flashed across the sea—even before that of the Queen—had been one of religious exultation. It was from the Directors in Great Britain to those on this side the Atlantic, and, simply reciting the fact that Europe and America were united by telegraph, at once broke into a strain of religious rapture, echoing the song of the angels over a Saviour's birth: "Glory to God in the highest; on earth, peace, good-will toward men." Poetry at once caught up the strain. The event became the theme of innumerable

acquainted, and well deserves the title prefixed to it by the laborious compiler:

"THE ATLANTIC TELEGRAPH.—Its Origin and History, with an Account of the Voyages of the Steamers Niagara and Agamemnon, in Laying the Cable, and of the Celebration of the Great Event in New-York, Philadelphia, Brooklyn, Montreal, Dublin, Paris, etc.; together with the Discussions, Sermons, Poetry, and Anecdotes relating thereto; also, a History of the Invention of the Electric Telegraph. Illustrated with Maps, Plans, Views, and Portraits, collected from the Newspapers of the Day, and arranged by John Russell Bartlett. 1858."

odes and hymns, of which it must be said that, whatever their merit as poetry, their spirit at least was noble, celebrating the event chiefly as promoting the brotherhood of the human family. The key-note was struck in such lines as these:

>'Tis done! the angry sea consents,
> The nations stand no more apart,
>With claspèd hands the continents
> Feel throbbings of each other's heart.
>
>Speed, speed the cable; let it run
> A loving girdle round the earth,
>Till all the nations 'neath the sun
> Shall be as brothers of one hearth;
>
>As brothers pledging, hand in hand,
> One freedom for the world abroad,
>One commerce over every land,
> One language and one God.

The sermons preached on this occasion were literally without number. Enough found their way into print to make a large volume. Never had an event touched more deeply the spirit of religious enthusiasm. Devout men held it as an advance toward that millennial era which was at once the object of their faith and hope. Was not this the predicted time when, "many should run to and fro, and knowledge should be increased?" So said the preachers, taking for their favorite texts the vision of the Psalmist, "Their line is gone out through all the earth, and their words to the end of the world;" or the question of Job: "Canst thou send forth the lightnings, that they may go and say unto thee, Here we are?" Was not this the dawn of that happy age, when all men should be bound together in peaceful intercourse, and nations should learn war no more? Such was the burden of the discourses that were preached in a thousand pulpits from one end of the country to the other. Even the Roman Catholic Church, so lofty and inflexible in its claims, soaring into the past centuries, and almost disdaining the material progress of the present day as compared with the spiritual glories of the Ages of Faith, did not ignore the great event; and in laying the foundation of the new Cathedral of St. Patrick, the largest temple of religion on the continent, Archbishop Hughes placed under the corner-stone an inscription, wherein, along with the enduring record of the Christian faith and the names of martyrs and confessors, he did not disdain to include a brief memorial of this last achievement of science, and the name of him who had conferred so great a benefit on mankind.

These public demonstrations culminated on the first of September, when the city authorities gave a public ovation to Mr. Field and the officers of the expedition. In accepting these honors, Mr. Field had taken good care that the British officers should be included with the American. At St. John's he had been notified of the intended celebration, and at once telegraphed to the British Admiral at Halifax:

"I should consider it a very great personal favor if you would permit the Gorgon, Captain Dayman, to accompany the Niagara, Captain Hudson, to New-York. English officers and English sailors have labored with American officers and American sailors to lay the Atlantic cable. They were with us in our days of trial, and pray let them, if you can, share with us our triumph."

The request was granted so far as this, that the officers were allowed leave of absence, and came on to New-York to take part in the celebration, and in all the honors which followed, the officers of the Gorgon were associated with those of the Niagara.

The day arrived, and the celebration surpassed any thing which the city had ever witnessed before. It was a mild autumn day—warm, yet with a sky softly veiled with clouds, that seemed to invite a whole population into the streets. The day commenced with a solemn service at Trinity Church, which was attended by the city authorities, the representatives of foreign powers, and an immense concourse of people. The vast edifice was decorated with evergreens; in the centre hung a cross, with the inscription: "Glory to God on high; and on earth, peace, good will towards men."

When the audience were assembled, there entered a procession of two hundred clergy, headed by Bishop Doane of New-Jersey, who was to deliver the address. Prayers were offered and Scriptures were read, and at intervals the choir burst forth in those anthems in which for ages the Church has been wont to pour forth its joy and exultation: "O come, let us sing unto the Lord," the *Gloria in Excelsis*, and the *Te Deum Laudamus*.

At noon, Mr. Field and the officers of the ships landed at Castle Garden and were received with a national salute. A procession was formed which extended for miles through the streets from the Battery to the Crystal Palace. In the procession were Lord Napier, the British Minister, and officers of the army and navy. For the whole distance the streets were crowded. The windows and even the tops of the houses were filled with people. Everywhere flags and banners, with every device, floated in the air. So dense was the crowd that it was five or six hours before the procession could reach the Crystal Palace.

Here its coming was awaited by an assembly that filled all the aisles and galleries. An address was delivered, giving the history of the Atlantic Telegraph. The Mayor then rose, and presenting Mr. Field to the audience, spoke as follows:

"SIR: History records but few enterprises of such

'great pith and moment' as to command the attention and at the same time enlist the sympathies of all mankind. In all ages warlike expeditions have been undertaken on a scale of grandeur sufficient to astonish the world; but the evils which are inseparable from their prosecution have always sent a thrill of horror through the anxious nations. The discovery of the Western continent even, the grandest event of modern times, was made by an insignificant fleet which left the shores of Portugal without attracting the notice of the civilized world. Far different has been the history of the daring and difficult enterprise of uniting the Old World and the New by means of the electric telegraph. From the very outset the good, the great and the wise of all lands beneath the sun, have watched with intense anxiety, and even when doubt existed, with warm interest, every step taken toward the accomplishment of what was universally acknowledged to be the most momentous undertaking of an age made marvellous by wonderful scientific and mechanical achievements. The two greatest and freest nations of the globe, by independent constitutional legislation, and by the aid of their finest ships and their ablest officers and engineers, combined together to insure success. Capital was liberally subscribed by private citizens in a spirit which put greed to the blush. The press on both sides of the Atlantic recorded the details of the progress of the undertaking with cordial interest, and secured the generous sympathies of men of all kindreds and tongues and nations in its behalf. You were thus fortunate, sir, in being identified with a project of such magnificent proportions and universal concern. But the enterprise itself was no less fortunate in being projected and carried into execution by a man whom no obstacles could daunt, no disasters discourage, no doubts paralyze, no opposition dishearten. If you, to whom the conduct of this great enterprise was assigned by the will of Providence and the judgment of your fellow-men, had been found wanting in courage, in energy, in determination, and in a faith that was truly sublime, the very grandeur of the undertaking would only have rendered its failure the more conspicuous. But, sir, the incidents of the expedition, and the final result—too familiar to all the world to need repetition here—have demonstrated that you possessed all the qualities essential to achieve a successful issue. It is for this reason that you now stand out from among your fellow-men a mark for their cordial admiration and grateful applause. The city of your home delights to honor you; your fellow-citizens, conscious that the glory of your success is reflected back upon them, are proud that your lot has been cast among them. They have already testified their appreciation of your great services and heroic perseverance by illuminations, processions, serenades, and addresses. And now, sir, the municipal government of this, the first city on the Western continent, instruct me, who have never felt the honor of being its chief magistrate so sensibly as in the presence of this vast assemblage of its fair women and substantial citizens, to present to you a gold box, with the arms of the city engraved thereon, in testimony of the fact that to you mainly, under Divine Providence, the world is indebted for the successful execution of the grandest enterprise of our day and generation; and in behalf of the Mayor, Aldermen, and Commonalty of the city of New-York, I now request your acceptance of this token of their approbation. In conclusion, sir, of this, the most agreeable duty of my public life, I sincerely trust that your days may be long in the land, and as prosperous and honorable as your achievement in uniting the two hemispheres by a chord of electric communication has been successful and glorious."

To this flattering address, Mr. Field replied:

"SIR: This will be a memorable day in my life; not only because it celebrates the success of an achievement with which my name is connected, but because the honor comes from the city of my home — the metropolitan city of the new world. I see here not only the civic authorities and citizens at large, but my own personal friends—men with whom I have been connected in business and friendly intercourse for the greater part of my life. Five weeks ago, this day and hour, I was standing on the deck of the Niagara, in mid-ocean, with the Gorgon and Valorous in sight, waiting for the Agamemnon. The day was cold and cheerless, the air was misty, and the wind roughened the sea; and when I thought of all that we had passed through—of the hopes thus far disappointed, of the friends saddened by our reverses, of the few that remained to sustain us—I felt a load at my heart almost too heavy to bear, though my confidence was firm, and my determination fixed. How different is the scene now before me—this vast crowd testifying their sympathy and approval, praises without stint, and friends without number! This occasion, sir, gives me the opportunity to express my thanks for the enthusiastic reception which I have received, and I here make my acknowledgments before this vast concourse of my fellow-citizens. To the ladies I may, perhaps, add, that they have had their appropriate place, for when the cable was laid, the first public message that passed over it came from one of their own sex. This box, sir, which I have the honor to receive from your hand, shall testify to me and to my children what my own city thinks of my acts. For your kindness, sir, expressed in such flattering, too flattering terms, and for the kindness of my fellow-citizens, I repeat my most heartfelt thanks."

The enthusiasm with which this address was received reached its height, when at the close, Mr. Field advanced to the edge of the platform, and unrolling a despatch, held it up, saying: "Gentlemen, I have just received a telegraphic message from a little village, now a suburb of New-York, which I will read to you:

"LONDON, September 1, 1858.

"To CYRUS W. FIELD, New-York:

"The directors are on their way to Valentia, to make arrangements for opening the line to the public. They convey, through the cable, to you and your fellow-citizens, their hearty congratulations and good wishes, and cordially sympathize in your joyous celebration of the great international work."*

A gold medal was presented to Captain Hudson, with an address, to which he made a fitting reply. Similar testimonials were presented to all the English captains through Mr. Archibald, the British Consul,

* The history of this despatch is curious. Though dated at London, it was sent from a small town in Ireland. The directors were on their way from Dublin to Valentia, on the morning of the first of September, when Mr. Saward remarked: "This is the day of the celebration in New-York—we ought to send a despatch to Mr. Field." Accordingly, at the first stopping-place, (we think it was Mallow Station,) the message was written, and forwarded to Valentia, and thence sent across the Atlantic. It was put into Mr. Field's hand just as he was getting into his carriage on the Battery.

who replied for his absent countrymen, after which the whole audience rose to their feet, as the band played "God save the Queen."

It was long after dark when the exercises closed, and the vast multitude dispersed.

The night witnessed one of those displays for which New-York surpasses all the cities of the world—a firemen's torchlight procession—a display such as was afterward given to the Prince of Wales, but which we shall probably witness no more, since the Volunteer Fire Department is disbanded.

But one day did not exhaust the public enthusiasm. The next evening, a grand banquet was given by the city authorities, at which were present a great number of distinguished guests. Lord Napier spoke, in language as happy as it was eloquent, of the new tie that was formed between kindred dwelling on opposite sides of the sea, and awarded the highest praise to the one whom he recognized as the author of this great achievement. Mr. Field replied, modestly disclaiming the "too much honor" that was heaped upon him, saying that it did not belong to him alone, and seemed most anxious to do full justice to all, on both sides of the Atlantic, who had shared in the great work.

Of course, we have no wish to recall these faded festivities, or to rehearse all the sentiments and speeches of that night of rejoicing. It is beyond the power of any artist to reproduce such a carnival, for he cannot put on canvas the spirit of the scene. Even in trying to recall it now, we feel

> Like one who treads alone
> Some banquet-hall deserted.

Since then, years have passed, and the shouts and cheers that rang in that hall have died away to an echo. Speakers and actors, many of them, have passed from the earth. Happy is it if we may say, that the work which they celebrated, remains.

While these demonstrations continued, every opposing voice was hushed in the chorus of national rejoicing; yet some there were, no doubt, who looked on with silent envy or whispered detraction. But who could grudge these honors to the hero of the hour— honors so hardly won, and which, as it proved, were soon to give place to harsh censures and unjust imputations?

Alas for all human glory! Its paths lead but to the grave. Death is the end of human ambition. That very day that a whole city rose up to do honor to the Atlantic Telegraph and its author, it gave its last throb, and that first cable was thenceforth to sleep for ever silent in its ocean grave.

CHAPTER XII.

SUDDEN STOPPAGE OF THE CABLE. REACTION OF PUBLIC FEELING. GRAVE SUSPICIONS OF BAD FAITH. DID THE CABLE EVER WORK? DECISIVE PROOF.

THE ATLANTIC CABLE WAS DEAD! That word fell heavy as a stone on the hearts of those who had staked so much upon it. What a bitter disappointment to their hopes! In all the experience of life there are no sadder moments than those in which, after years of anxious toil, striving for a great object, and after one supreme moment of complete success, the fruit of all these labors becomes a total wreck. Vain is all human toil and endeavor. The years thus spent are fled away; the labor that was to have borne such rich fruits of glory, is lost; and the prolonged tension of the mind by the excitement of hope and ambition, and the momentary dream of success, reäcts to plunge it into a deeper depression. So was it here. Years of labor and millions of capital were swept away in an hour into the bosom of the pitiless sea.

Of course the reäction of the public mind was very great. As its elation had been so extravagant before, it was now silent and almost sullen. People were ashamed of their late enthusiasm, and disposed to revenge themselves on those who had been the objects of their idolatry. It is instructive to read the papers of the day. As soon as it was evident that the Atlantic cable was a dead lion, many hastened to give it a parting kick. There was no longer any dispute as to who was the author of the great achievement. Rival claimants quietly withdrew from the field, content to leave him "alone in his glory."

Many explanations were offered of this sudden suspension of life. One writer argued that the Telegraphic Plateau was only a myth; that the bottom of the ocean was jagged and precipitous; and that the cable passed over lofty mountain chains, and had hung suspended from the peaks of submarine Alps, till it broke and fell into the tremendous depths below.

But others found a readier explanation. With the natural tendency of a popular excitement to rush from one extreme to the other, many now believed that the whole thing was an imposition on public credulity, a sort of "Moon hoax" or a gigantic speculation. An elaborate article appeared in a Boston paper, headed with the alarming question, "Was the Atlantic cable a humbug?" wherein the writer argued through several columns that it was a huge deception. A writer in an English paper also made merry of the celebration in Dublin, where a banquet was given to Sir Charles Bright, in an article bearing the ominous title: "Very like a whale!" This writer proved not only that the Atlantic cable was never laid, but that such a thing was mathematically impossible. But since he turned out to be a crazy fellow, whom the police had to take into custody, his "demonstrations" did not make much impression on the public. The difficulty of finding a *motive* for the perpetration of such a stupendous fraud, did not at all embarrass these ingenious writers. Was it not enough to make the world stare? to furnish something to the gaping crowd, even though it were but a nine days' wonder? Those who thus reasoned seemed not to reflect that such deceptions are always sure to be found out; that one who goes up like a rocket may come down like a stick; and that if by false means he has made himself an object of popular idolatry, he is likely to become the object of popular indignation.

But others there were—sharp, shrewd men—who thought they could see through a mill stone farther than their neighbors, who shook their heads with a knowing air, and said: "It was all a stock speculation." One writer stepped before the public with this solemn inquiry: "Now that the great cable glorification is over, we should like to ask one question, How many shares of his stock did Mr. Field sell during the month of August?" This he evidently thought was a question which could not be answered, except by acknowledging a great imposition on the public. If this brilliant inquirer after truth really desired to be informed, we could have referred him to Messrs. George Peabody & Co., of London, with whom was deposited all of Mr. Field's stock at the time, and who, during that memorable month of August, sold *just one share*, and that at a price below the par value, which was paid by Mr. Field himself. Whether this was an object sufficiently great to set two hemispheres in a blaze, we leave him to judge.

To those who have followed this narrative, all these conjectures and suspicions will appear very absurd. These personal reflections we would treat with contempt, as a man of character always scorns an imputation on his personal honor. But while we despise these anonymous scribblers, as they deserve, yet we recognize the fact that many honest people not disposed to think evil were sorely perplexed. That the cable should continue to work for three or four weeks, *and then stop the very day of the celebration*, was a circumstance certainly very singular, if not suspicious; and it was not to be wondered at that it should excite a painful feeling of doubt. This distrust is quite natural, and ought not to be matter either of offence or surprise. On the contrary, those who are fully satisfied of the facts, ought rather to be glad of the opportunity which such questions afford, to present in full the amplest vindication.

To answer all inquiries, we propose to give a very brief history of the working of the Atlantic cable. It was landed on both sides of the ocean on the fifth of August. The last recorded message passed over it on the first of September, one day short of four weeks. Within that time there were sent exactly four hundred messages, of which two hundred and seventy-one were from Newfoundland to Ireland, and one hundred and twenty-nine from Ireland to Newfoundland. Of these, the greater part were merely between the operators themselves, respecting the adjustment of instruments, and working the telegraph, which, while they furnished decisive evidence *to them*, are of no force to the public. Of course an operator, working with a battery on the shore at Valentia, or at Trinity Bay, watching his instrument, and seeing the spark of light, needs no other evidence of an electric current that has passed through the cable. He *sees* it, and knows, as if he saw the flash of a gun on the coast of Ireland, that it is a light which has come from beyond the sea. When he hears the familiar click, he knows that it is a voice whispering to him out of the bosom of the waters. But these are of no value to the public as deciding the fact of actual communication. What they need is *public* messages, conveying news from one hemisphere to the other. Of these, there were not a great number, for obvious reasons. The cable, during the four weeks of its existence, never worked *perfectly*—that is, as a land line works, transmitting messages freely and rapidly, and with perfect accuracy. It worked, but slowly, and with frequent interruptions, for reasons which we will state, and which, we think, will satisfy any one that the wonder is, not that it did so little, but that it did so much.

1. To begin with, the cable was not constructed in the most perfect manner. Its makers, though the best then in the world, had had but little experience in making deep-sea cables. No line over three hundred miles long had ever been laid. 2. It had been made more than a year before. After it was finished, part of it had been coiled out of doors, where it was exposed to a burning sun, by which, as was afterward found, the gutta-percha had been melted in many places till the insulation was nearly destroyed. 3. It had been put on board the ships in 1857, and after the first failure, had been taken out and coiled on the dock at Plymouth, and then re-shipped in 1858. Thus it had been twisted and untwisted, some portions of it as many as ten times. Then the half on board the Agamemnon was so shaken in the terrible gale of June, that it was seriously injured, and some portions were

cut out and condemned. Take all these things together, and the wonder is, not that the cable failed after a month, *but that it ever worked at all.*

Owing to this impaired state of the cable, we admit fully that it did *not* work perfectly. Signals came and went, which showed that the electric current passed freely from shore to shore, and gave promise that with more delicate instruments it could be taught to speak plainly. But for the present it spoke slowly and with difficulty. It often took hours to get through a single despatch, if of any length. Witness the delay in transmitting the Queen's message. These frequent interruptions were ascribed to various causes. Sometimes it was earth-currents; at others, a thunderstorm was raging. Thus, on the morning of Thursday, the twenty-sixth of August, there was a violent storm in Newfoundland, heavy rain, accompanied by thunder and lightning. At three o'clock, the lightning was so intense that for about an hour and a half the end of the cable had to be put to the earth for protection. After that the storm cleared away, and at seven o'clock the weather is reported as being very fine. But aside from these local and temporary causes, the real difficulty was in the cable itself, whose insulation had been fatally impaired, and which was now wearing out its life on the rocks of the sea. These causes made its speech difficult and broken. Yet sometimes it flashed up with sudden power. Thus, in one case, a message was sent from the office at Trinity Bay to Ireland and an answer received back *in two minutes*. Such incidents excited the liveliest hopes that all difficulties would be speedily overcome, and justified the messages which were sent to the New-York papers from day to day, that the instruments were being adjusted by which it was expected that the line would soon be put in perfect working order, and be thrown open to the public. But these flashes of light proved to be only the flickering of the flame, that was soon to be extinguished in the eternal darkness of the waters.

But the question which the enemies of the Atlantic Telegraph have chosen to raise is, not whether the cable worked fast or slow, *but whether it ever worked at all.* Happily, this is a question which can easily be settled, since it is one simply of facts and dates, which can be ascertained by referring to the files of the English and American papers. Of course what we ask in this case is messages containing *news*. Mere congratulations between the Queen and the President, or the Mayor of New-York and the Mayor of London, prove nothing, for these might be prepared beforehand, if we suppose a design to impose on the credulity of the public. But the decisive test is this: Was there at any time within that month published in the English or American journals NEWS which could not be matter of guess or conjecture, and within a time too short for its possible transmission in any other way? If this can be proved beyond all doubt, even in a few instances, the question is decided, for the argument is just as strong with a dozen cases as with a thousand. We give, therefore, a few dates, the accuracy of which can be tested by any one who will take the trouble to examine the English and American papers:

I. On Saturday, the fourteenth of August, the steamships Arabia and Europa, the former bound for New-York and the latter for Liverpool, came into collision off Cape Race. The accident was not known in this city until Tuesday, the seventeenth, since it could not be telegraphed here till the Arabia reached Halifax or the Europa St. John's, into which port she put for repairs. As soon as the news reached New-York, the Agent of the Company, Mr. Nimmo, (Mr. Cunard himself being then in England,) at once prepared a despatch to be sent to relieve immediate anxiety. This was not forwarded to Newfoundland, as peremptory orders had been given not to transmit any private business messages to go through the cable until the line was fully open to the public. But the next day Mr. Field arrived in New-York, and Mr. Nimmo applied to him. Seeing the urgency of the case, he ordered it to be forwarded. It was accordingly sent, and arrived in London on the twentieth, giving the first news that was received of the accident. This was repeatedly stated by the late Sir Samuel Cunard, of London, and is confirmed by Mr. Edward Cunard, of New York. The message was published in the London papers of the twenty-first, and is as follows:

"Arabia in collision with Europa, Cape Race, Saturday. Arabia on her way. Head slightly injured. Europa lost bowsprit, cutwater; stem sprung. Will remain in St. John's ten days from sixteenth. Persia calls at St. John's for mails and passengers. No loss of life or limb."

This first news message was not only a very decisive one as to the *fact* of telegraphic communication, but one which showed the *benefits* which it would confer. Mr. William E. Dodge, a well-known merchant of New-York, says: "I was in Liverpool at the time, and expecting friends by the Europa. Any delay in the arrival of the ship would have caused great anxiety. But one morning, on going down to the Exchange, we saw posted up this despatch received the night before by the Atlantic Telegraph. All then said if the cable never did any thing more, it had fully repaid its cost." Well may he add with devout feeling: "It seemed as if Divine Providence had permitted the event, to furnish a testimony which could not be denied, to the reality and the benefit of this new means of communication between the two continents."

II. Passing over all the messages exchanged between the operators at the stations, the congratulations of Queen and President, and of the Mayors of New-York and London, we come to another news despatch; August twenty-fifth, Newfoundland reports to Valentia:

"Persia takes Europa's passengers and mails. Great rejoicing everywhere at success of cable. Bonfires, fireworks, *feux de joie*, speeches, balls, etc., Mr. Eddy, the first and best telegrapher in the States, died to-day. Pray give some news for New-York; they are mad for news."

In the above despatch, we remark especially one item, the death of Mr. Eddy, an announcement which the writer, who was then in Europe, read first in the London Times, and which arrested his attention, as he had some acquaintance with that gentleman.* Those

* Mr. James Eddy died suddenly, at Burlington, Vermont, Monday, August twenty-third, 1858, at ten o'clock, fifteen minutes A.M. The exact day and hour we learn from his widow, who is now living in Brooklyn. The news was telegraphed to New-York, and from there sent to Trinity Bay, from which it was forwarded to Valentia, and appeared in the London Times Wednesday morning. Thus not forty-eight hours had elapsed after he breathed his last, before it was published in England. If any one wishes to see the despatch, he will find a file of the London Times in the Astor Library.

P. S.—Slight discrepancies are sometimes the strongest possible confirmation of truth, as they show that there was no thought of imposition. One of these appears here. The despatch is dated August twenty-fifth, and says Mr. Eddy died *to-day*, and yet it is published who argue so strenuously for the theory of collusion and deception, must be somewhat embarrassed to account for this. Do they suppose that this death was a matter of concert and design? that Mr. Eddy died on that day, so that a message, which they must assume to have been sent two weeks beforehand, could be proved correct? This is an absurdity too gross even for them, yet to such absurdities are they reduced by denial that authentic messages ever passed over the Atlantic cable.

To the demand for news in the above despatch, a reply was at once returned: "Sent to London for news." And later the same day came the following: "North-American with Canadian, and the Asia with direct Boston mails, leave Liverpool, and Fulton, Southampton, Saturday next. To-day's morning papers have long, interesting reports by Bright. Indian news. Virago arrived at Liverpool to-day; Bombay dates nineteenth July. Mutiny being rapidly quelled."

A despatch of the same date, August 25th, also announces peace with China. The whole was received at Trinity Bay about nine o'clock P.M., and in the London Times of the same date. How is this? It was sent between nine and ten o'clock at night of the twenty-fourth, when the operator would say *this day* of a piece of news just received, but in affixing the date, he was governed *by Greenwich time*, which made it more than three hours later. Accordingly it is published in the London Times, dated August twenty-fifth, fifty-three minutes past twelve A.M.

would have been sent on at once to New-York, but that the land lines in Nova Scotia were closed at that hour. It was sent the next morning, and appeared in the evening papers of the twenty-sixth.

By referring again to the London Times, the reader will see that the news from China was published in London on the twenty-third of August. It is there given as *very unexpected news*, so that it could not have been a shrewd guess on the part of any body either in England or America. It took the public by surprise, both for the news itself and *for the way in which it came*—which was not by India and the Red Sea, but by St. Petersburg, where it arrived on the twenty-first, having been brought overland by a courier to Prince Gortchakoff. From there it was telegraphed to the Government at Paris, and thence to London. The Times comments on this roundabout way in which intelligence so important reached England. Yet this news, so unlooked for, announced in London only on the morning of the twenty-third of August, was published in New-York on the twenty-sixth.

III. August twenty-seventh, comes a still longer despatch, which we give in full: " George Saward, Secretary Atlantic Telegraph Company, to Associated Press, New-York. News for America by Atlantic cable. Emperor of France returned to Paris, Saturday. King of Prussia too ill to visit Queen Victoria. Her Majesty returns to England thirtieth of August.—St. Petersburg, twenty-first of August. Settlement of Chinese question. Chinese empire opened to trade; Christian religion allowed; foreign diplomatic agents admitted; indemnity to England and France.—Alexandria, August ninth. The Madras arrived at Suez seventh inst. Dates Bombay to the nineteenth; Aden, thirty-first. Gwalior insurgent army broken up. All India becoming tranquil."

This despatch embodies about a dozen distinct items of news, not one of which could be known without a telegraphic communication. The whole was received in New-York, and published in the evening papers *the same day.*

IV. Not to be outdone in giving news, the next day, Saturday, August twenty-eighth, Newfoundland thus replies to Valentia:

"TO THE DIRECTORS: Take news first, Saward. Sir William Williams, of Kars, arrived Halifax Tuesday. Enthusiastically received. Immense procession—welcome address—feeling reply. Held levee—large number presented. Niagara sailed for Liverpool at one this morning. The Gorgon arrived at Halifax last night. Yellow fever in New-Orleans, sixty to seventy deaths per day. Also declared epidemic, Charleston. Great preparations in New-York and other places for celebration, to be held the first and second of

September. New-Yorkers will make it the greatest gala-day ever known in this country. Hermann sailed for Fraser's River; six hundred passengers. Prince Albert sailed yesterday for Galway. Arabia and Ariel arrived New-York; Anglo Saxon, Quebec; Canada, Boston. Europa left St. John's this morning. Splendid aurora Bay of Bulls to-night, extending over eighty-five degrees of the horizon."

Let any one examine carefully this despatch, considering the minuteness of the details—which could not be known or conjectured—such as the appearance of yellow fever at New-Orleans, with the number of deaths a day; the sailing or arrival of seven steamers; the number of passengers for Fraser's River, etc.—and then examine the London Times, and see that all these items appeared in it Monday morning, August thirtieth, and if he does not admit that collusion or deception is out of the question, no amount of evidence could convince him.

V. We will give but one proof more. On the last day of August, the day before the cable ceased to work, Valentia sent to Newfoundland two messages for the British Government, both signed by "the Military Secretary to the Commander-in-Chief, Horse Guards, London," and addressed—the first to General Trollope, Halifax, which said, "The sixty-second regiment is not to return to England;" and the other to the General Officer commanding at Montreal, saying: "The Thirty-ninth regiment is not to return to England." This was the time of the Sepoy rebellion, and the Government had sent out orders by mail for these two regiments to embark immediately for home, to be sent to India. But the mutiny being nearly suppressed, this was found not to be necessary, and the prompt countermanding of the order by telegraph, saved the British Government, in the cost of transportation of troops, not less than fifty thousand pounds. The despatch to Halifax was received the same day that it was sent from London. The sending of this despatch, and its almost immediate reception, is attested by an official letter from the War Office in London.

But why multiply arguments? The facts here given are accessible to all who have the candor and the patience to examine. Let any man take the files of English and American papers issued during that month of August, and compare them day by day, and if he is not thoroughly satisfied that a great number of authentic messages passed over the Atlantic cable, he is beyond the reach of human testimony. His case is one of "invincible ignorance." Neither would he be persuaded though one rose from the dead.

CHAPTER XIII.

Attempts to Revive the Company—Profound Discouragement—It applies to the Government for Aid, which declines to give an unconditional Guarantee—Failure of the Red Sea Telegraph—Scientific Experiments—Cables laid in the Mediterranean and the Persian Gulf—Efforts to raise Capital in the United States and in England—Brief History of the next Five Years.

It takes a long time to recover from a great disaster. When at last the friends of the Atlantic Telegraph were obliged to confess that the cable had ceased to work; when all the efforts of the electricians failed to draw more than a few faint whispers, a dying gasp, from the depths of the sea, there ensued in the public mind a feeling of profound discouragement. For a time this paralyzed all effort to revive the Company and to renew the enterprise. And yet the feeling, though natural, was extreme. If they had not done all they attempted, they had accomplished much. They had at least demonstrated the possibility of laying a cable across the Atlantic Ocean, and of sending messages through it. This alone was no small triumph. So men reasoned when sober reflection returned, and at length the tide of public confidence, which had ebbed so strongly, began to reflow, and once more to creep up the shores of England.

But when a great enterprise has been overthrown, and lies prostrate on the earth, the first impulse of its friends is to call on Cæsar for help. So the first appeal of the Atlantic Telegraph Company was to the British Government. It was claimed, and with reason, that the work was too great to be undertaken by private capital alone. It was a matter, not of private speculation, but of public and national concern. It was, therefore, an object which might justly be undertaken by a powerful government, in the interest of science and of civilization.

To raise capital for a new cable, it was necessary to have some better security than the hazards of a vast and doubtful undertaking. Hence the Company asked the Government to guarantee the interest on a certain amount of stock, even if the second attempt should not prove a complete success. With such a guarantee, the capital could be raised in London in a day.

In this application they might have been successful, but for an untoward event, which dampened the confidence of the public in all submarine enterprises—the failure of the Red Sea Telegraph. The British Government, anxious to forward communication with India, had given that Company an unconditional guarantee, on the strength of which the capital was raised, and the cable manufactured and laid. But in a very short time it ceased to work. This proved a serious loss to the treasury of Great Britain. To the public, which did not understand the cause of the failure to be the imperfect construction of the cable, the effect was to impair confidence in all long submarine telegraphs. Of course, after such an experience, the Government was not disposed to bind itself by such pledges again. It was, however, ready to aid the enterprise by any safe means. It therefore increased its subsidy from fourteen thousand pounds to twenty thousand pounds; and guaranteed eight per cent on six hundred thousand pounds of new capital for twenty-five years, with only one condition—*that the cable should work.* This was a liberal grant, and under the circumstances, was all that could be expected.

Still further to encourage the undertaking, it ordered new soundings to be taken off the coast of Ireland. These were made by Captain Hoskins, R.N., and dispelled the fears which had been entertained of a submarine mountain, which would prove an impassable barrier in the path of an ocean telegraph.

But the greatest service which the British Government rendered, was in the long course of experiments which it now ordered, to determine all the difficult problems of submarine telegraphy. In 1859, the year after the failure of the first Atlantic cable, the Board of Trade appointed a committee of the most eminent scientific and engineering authorities in Great Britain to investigate the whole subject. This was composed of Captain Douglas Galton, of the Royal Engineers, now of the War Office, London, who specially represented the Government; Professor Wheatstone, the celebrated electrician; William Fairbairn, late President of the British Association for the Advancement of Science; George Parker Bidder, whose name ranks with those of Stephenson and Brunel; C. F. Varley, who, in the practical working of telegraphs, has no superior in England; Latimer Clark and Edwin Clark, both engineers, who had had great experience in the business of telegraphing; and George Saward, the Secretary of the Atlantic Telegraph Company.

This Committee sat for nearly two years, at the end of which, it made a report to the Government, which fills a very large volume, in which are detailed an immense number of experiments, touching the form and size of cables, their relative strength and flexibility, the power of telegraphing at long distances, the speed at which messages could be sent; and in fine, every possible question, either as to the electrical or engineering difficulties to be overcome. The result of these manifold and laborious experiments is summed up in the following certificate, signed by all who had taken part in this memorable investigation:

"LONDON, 13th July, 1863.

"We, the undersigned, members of the Committee, who were appointed by the Board of Trade, in 1859, to investigate the question of submarine telegraphy, and whose investigation continued from that time to April, 1861, do hereby state, as the result of our deliberations, that a well-insulated cable, properly protected, of suitable specific gravity, made with care, and tested under water throughout its progress with the best known apparatus, and paid into the ocean with the most improved machinery, possesses every prospect of not only being successfully laid in the first instance, but may reasonably be relied upon to continue for many years in an efficient state for the transmission of signals.

DOUGLAS GALTON,   CROMWELL F. VARLEY,
C. WHEATSTONE,   LATIMER CLARK,
WM. FAIRBAIRN,   EDWIN CLARK,
GEO. P. BIDDER,   GEO. SAWARD."

Thus the years which followed the failure of 1858—though they saw no attempt to lay another ocean cable—were not years of idleness. They were rather years of experiment and of preparation, clearing the way for new efforts and final victory. The Atlantic Telegraph itself had been a grand experiment. It had taught many important truths which could be learned in no other way. Not only had it demonstrated the possibility of telegraphing from continent to continent, but it had been useful *even in exposing its own defects*, as it taught how to avoid them in the future.

For example, in working the first cable, the electricians had thought it necessary to use an enormous battery. They did not suppose they could reach across the whole breadth of the Atlantic, and touch the Western hemisphere, unless they sent an electric current that was almost like a stroke of lightning; and that, in fact, endangered the safety of the conducting wire. But they soon found that this was unnecessary. God was not in the whirlwind, but in the still, small voice. A soft touch could send a thrill along that iron nerve. It seemed as if the deep were a vast whispering gallery, and that a gentle voice murmured in the ocean caves, like a whisper in a sea-shell, might be caught, so wonderful are the harmonies of nature, by listening ears on remote continents. Thus was given a new meaning to the poet's

> "Airy tongues, that syllable men's names
> On sands, and shores, and desert wildernesses."

These were also years of great progress, not only in the science of submarine telegraphy, but in the construction of deep-sea cables. In spite of the failure of that in the Red Sea, one was laid down in the Mediterranean, fifteen hundred and thirty-five miles long, from Malta to Alexandria, and another in the Persian

Gulf, fourteen hundred miles long, by which telegraphic communication was finally opened from England to India. Both these lines still continue in perfect working order. Others were laid in different seas and oceans in distant parts of the world. These great triumphs, following the scientific experiments which had been made, revived public confidence, and prepared the way for a fresh attempt to pass the Atlantic.

Yet not much was done to renew the enterprise until 1862. Mr. Field had been indefatigable in his efforts to reanimate the Company. He was continually going back and forth to the British Provinces and to England, urging it wherever his voice could be heard. Yet times were adverse. The United States had been suddenly involved in a tremendous war, which called into the field hundreds of thousands of men, and entailed a burden of many hundreds of millions. While engaged in this life-and-death struggle, and rolling up such a mountain of debt, our people had little thought to bestow on other great enterprises by land or sea.

And yet one incident of the war forcibly recalled public attention to the necessity of some speedier communication with Europe than by steam. The unhappy Trent affair aroused an angry feeling in Great Britain which nearly resulted in hostilities, all of which might have been prevented by a single word of explanation.

As the London Times said truly: "We nearly went to war with America because we had not a telegraph across the Atlantic." After such a warning, it was natural that both countries should begin to think seriously of the means of preventing future misunderstanding. Mr. Field went to Washington, and found great readiness on the part of the President and his Cabinet to encourage the enterprise. Mr. Seward wrote to our Minister in London that the American Government would be happy to join with that of Great Britain in promoting this international work. With this encouragement, Mr. Field went to England to urge the Company to renew the undertaking. While in London, he endeavored to obtain from some responsible parties an offer to construct and lay down a cable. Messrs. Glass, Elliot & Co. replied, declaring their willingness to undertake the work, without at first naming the precise terms. They wrote, under date of February seventeenth:

"SIR: In reply to your inquiries, we beg to state that we should not be willing to manufacture and lay a Submarine Telegraph Cable across the Atlantic, from Ireland to Newfoundland, assuming the entire risk, as we consider that would be too great a responsibility for any single firm to undertake; but we are so confident that these points can be connected by a good and durable cable, that we are willing to contract to do the work, and stake a large sum upon its successful laying and working.

"We shall be prepared in a few days, as soon as we can get the necessary information in regard to what price we can charter suitable ships for the service, to make you a definite offer."

Although it is anticipating a few months in time, we may give here the "definite offer," which was obtained by Mr. Field, on his return to England in the autumn. It was as follows:

"LONDON, October 20, 1862.

"CYRUS W. FIELD, ESQ., ATLANTIC TELEGRAPH COMPANY:

"DEAR SIR: In reply to your inquiries, we beg to state, that we are perfectly confident that a good and durable Submarine Cable can be laid from Ireland to Newfoundland, and are willing to undertake the contract upon the following conditions:

"First. That we shall be paid each week our actual disbursements for labor and material.

"Second. That when the cable is laid and in working order, we shall receive for our time, services, and profit twenty per cent on the actual cost of the line, in shares of the Company, deliverable to us, in twelve equal monthly instalments, at the end of each successive month whereat the cable shall be found in working order.

"We are so confident that this enterprise can be successfully carried out, that we will make a cash subscription for a sum of twenty-five thousand pounds sterling in the ordinary capital of the Company, and pay the calls on the same when made by the Company.

"Annexed we beg to hand you, for your guidance, a list of all the submarine telegraph cables manufactured and laid by our firm since we commenced this branch of our business, the whole mileage of which, with the exception of the short one between Liverpool and Holyhead, which has been taken up, is at this time in perfect and successful working order. The cable that we had the honor to contract for and lay down for the French Government, connecting France with Algeria, is submerged in water of nearly equal depths to any we should have to encounter between Ireland and Newfoundland.

"You will permit us to suggest that the shore ends of the Atlantic Cable should be composed of very heavy wires, as from our experience the only accidents that have arisen to any of the cables that we have laid have been caused by ships' anchors, and none of those laid out of anchorage ground have ever cost one shilling for repairs.

"The cable that we would suggest for the Atlantic will be an improvement on all those yet manfactured, and we firmly believe will be imperishable when once laid. We remain, sir, yours faithfully,

"GLASS, ELLIOT & Co."

The summer of this year Mr. Field spent in America, where he applied himself vigorously to raise capital for the new enterprise. To this end he visited different cities — Boston, Providence, Philadelphia, Albany, and Buffalo—to address meetings of merchants and others. All listened with interest, and applauded his courage and perseverance, and hoped he would succeed, but subscribed little. In New-York he succeeded better, but only by indefatigable exertions. He addressed the Chamber of Commerce, the Board of Brokers, and the Corn Exchange, and then he went almost literally from door to door, calling on merchants and bankers to enlist their aid. The result was, subscriptions amounting to about seventy thousand pounds, the whole of which was due to persevering personal solicitation. Even of those who subscribed, a large part did so more from sympathy and admiration of his indomitable spirit than from confidence in the success of the enterprise.

In England, however, the subject was better understood. For obvious reasons, the science of submarine telegraphy has made greater advances in that country than in ours. England is itself an island, and obliged to hold all its telegraphic communication with the continent by cables under the sea. Then it has large colonial possessions in all parts of the world, with which it is important to have the means of speedy communication. We can understand the pride of empire in a dominion on which the sun never sets—"a power," to quote the memorable description of our own Webster, "which has dotted the face of the whole globe with its possessions and military posts, whose morning drumbeat, following the sun and keeping company with the hours, encircles the whole earth with one continuous and unbroken strain of the martial airs of England." Was it strange that this mother of nations should reach out her long arms to embrace her distant children?

The pride of empire and necessity of her position had stimulated the spirit of enterprise in this direction. Hence it was that the subject of submarine telegraphs was so much better understood in England than in this country, not only by scientific men, but by capitalists. The appeal could be made to them with more assurance of intelligent sympathy. And yet so vast was the undertaking, that it required ceaseless effort to roll the stone to the top of the mountain, and the result was not completely achieved till the beginning of the year 1864.

CHAPTER XIV.

The Enterprise Renewed. Proposals for another Cable. Great Improvement on the Old. Generous Offer of the Manufacturers to take half the Capital. The Work begun. The Great Eastern and Captain Anderson. The whole Cable shipped on board in the Spring of 1865. Expedition in July.

It is a long night which has no morning. At last the day is breaking. While weary eyes are watching the East, "daylight comes over the sea." Five years have passed away, and though the time seemed long as an Arctic winter, that only made more bright the rising of the sun. Those years of patient experiment, when scientific men were applying tests without number, and submarine lines were feeling their way along the deep-sea floor in all the waters of the world, at last brought forth their fruit in that renewed confidence which is the forerunner of victory.

So strong was this feeling, that as early as August, 1863, although the capital was not raised, the Board advertised for proposals for a cable suitable to be laid across the Atlantic Ocean; and in order to leave invention entirely unfettered, they abstained from any dictation as to the form or materials to be adopted, merely stipulating for a working speed of eight words a minute.

In reply they received, in the course of a few weeks, seventeen different proposals from as many companies, many of them firms of large wealth and experience. These different tenders, with the numerous specimens of cable and materials, were at once submitted to the eminent Consulting Committee which had already rendered such service by its advice, embracing as it did the first engineering skill and scientific knowledge of the kingdom. This Committee examined all the proposals, and then, taking up one by one the different samples of cable, caused them in turn to be subjected to the severest tests. This took a long time, as it required a great number of experiments; but the result was highly satisfactory. The Committee were all of one mind, and recommended unanimously that the Board should accept the tender of Messrs. Glass, Elliot & Co., and the general principle of their proposed cable; but advised that before settling the final specification, every portion of the material to be employed should be tested with the greatest care, both separately and in combination, so as to ascertain what further improvements could be made. To this the manufacturers readily consented, feeling a noble ambition to justify the confidence of the Committee and the public. They provided abundant materials for fresh experiments. New cables were made and tested in different lengths; and experiments were also tried upon different qualities of wire and hemp, that were to compose its external protection. The result of all these investigations was the selection of a model which seemed to combine every excellence, and to approach absolute perfection.

Such was the cable which this eminent firm offered to manufacture, and to lay across the Atlantic, and that on terms so favorable, that it seemed as if it could not be difficult to raise the capital and proceed with the work. Indeed, a contract was partially made to that effect. So confident was Mr. Field, who was then in London, that an expedition would sail the following summer, that he insured his stock, part of it only against ordinary sea-risks, but part also to be laid and to work! But hardly had he left England before there was some unforeseen hitch in the arrangements, the money was not forthcoming, or some of the conditions were not complied with, and he had the mortification to receive letters, saying that the whole enterprise was postponed for another year!

This was indeed discouraging. Hope deferred maketh the heart sick, and this hope had been deferred for many a year. Yet this sudden dropping of the scheme did not imply a loss of interest or of faith on the part of those embarked in it. They believed in it as much as ever. But the general public did not respond to the call for more capital. Alas that the noblest enterprises should so often be delayed or defeated by the want of money! But so it is. Capital is always cautious and timid, and follows slowly in the path of great discoveries. While "star-eyed Science" flies far in advance of the human race, sordid Mammon creeps behind. If Columbus, instead of the patronage of Queen Isabella, had depended on a stock company for the means for his expedition, he might never have sailed from the shores of Spain. Happy was it for mankind that his faith and patience did not wear out, while going from court to court, and kingdom to kingdom, and almost begging his way from door to door!

But it is not in human nature—least of all in the Saxon blood—to despond long. The heart of man is like the sea, ebbing and returning with a ceaseless flow. Though at times it seems to have swept away to distant shores, yet as moons revolve and tides return, again the white-crested waves come rolling up the beach. Ten years of constant defeat would seem to have wrought a lasting discouragement. Yet again and again did the baffled spirit of enterprise return to the attempt. In January, 1864, Mr. Field was once more on his way to England, to try it again. He found the Directors, as before, deeply interested in the enterprise, and wishing it success. With a grateful heart he bore witness to their unfaltering courage. But mere courage and good wishes would not lay the Atlantic Telegraph. Yet what more could they do? They could not be expected to advance all the capital themselves. They had already subscribed liberally, and he could not ask them to do more. But with all the efforts that had been made in England and America, not half the capital was yet raised. The machinery was in a dead lock, with little prospect of being able to move. It was the misfortune of the enterprise that there was no one man who made it his sole and exclusive charge. The Board of Directors contained some of the best men in London. But they were, almost without exception, engaged in very large affairs of their own, with no leisure to make a public enterprise their special care. To insure success, it needed a trial of the one-man power—one brain, planning night and day; one agency incessantly at work, stirring up directors, contractors, and engineers; and one will pushing it forward by main strength. This was the force now to be applied.

The first element needed to put life into the old system was an infusion of new blood—new capital and new men. While the enterprise was in this state of collapse, Mr. Field addressed himself to a gentleman with whom, until then, he had no personal acquaintance, but who is well known in London as one of the largest capitalists of Great Britain—Mr. Thomas Brassey. Their first interview was somewhat remarkable. Referring to it a few months after, Mr. Field said: "When I arrived in this country, in January last, the Atlantic Telegraph Company trembled in the balance. We were in want of funds, and were in negotiations with the government, and making great exertions to raise the money. At this juncture I was introduced to a gentleman of great integrity and enterprise, who is well known, not only for his wealth, but for his foresight, and in attempting to enlist him in our cause he put me through such a cross-examination as I had never before experienced. I thought I was in the witness-box. He inquired of me the practicability of the scheme—what it would pay, and every thing else connected with it; but before I left him, I had the pleasure of hearing him say that it was a great national enterprise that ought to be carried out, and, he added, I will be one of ten to find the money required for it. From that day to this he has never hesitated about it, and when I mention his name, you will know him as a man whose word is as good as his bond, and as for his bond, there is no better in England."

Having thus secured one powerful ally, Mr. Field

took courage and went to work to make another. He says: "The words spoken by Mr. Brassey in the latter part of January, 'Let the Electric Telegraph be laid between England and America,' encouraged us all, and made us believe we should succeed in raising the necessary capital, and I then went to work to find nine other Thomas Brasseys, (I did not know whether he was an Englishman, a Scotchman, or an Irishman; but I made up my mind that he combines all the good qualities of every one of them,) and after considerable search I met with a rich friend from Manchester, [Mr. John Pender, M.P.,] and I asked him if he would second Mr. Brassey, and walked with him from 28 Pall Mall to the House of Commons, of which he is a member. Before we reached the House, he expressed his willingness to do so to an equal amount."

This was putting strong arms to the wheel. A few days after, a combination was formed to carry on the whole business of making Submarine Telegraphs, by a union of the Gutta-Percha Company with the firm of Glass, Elliot & Co., the principal manufacturers of sea cables, making one grand concern, to be called The Telegraph Construction and Maintenance Company. These two great capitalists entered into the new organization, of which Mr. Pender was made Chairman. The Gutta-Percha Company brought in still further strength to the joint enterprise, in the person of Mr. John Chatterton, and of Mr. Willoughby Smith, their electrician, and the inventor of the insulating material known as Chatterton's compound. The union of all these men made a combination of practical skill and financial ability, such as could be found in few companies in England or in the world. Mr. R. A. Glass was chosen Managing Director—a gentleman who seems born to be a manager, such power has he of gathering about him talent in every department and combining all into one complete organization. Thus reënforced by such powerful aid, the new Company now came forward, and offered at one stroke to take all the remaining stock of the Company. This was more than half the whole capital. As yet, of the £600,000 required, but £285,000 had been subscribed. Now this princely Company offered to take the balance themselves—£315,000. They did more—taking £100,000 of bonds beside. Thus at one dead lift these stalwart Englishmen took the whole enterprise on their broad shoulders. From that hour the problem was solved. Thus after a dead lock of six months the wheels were unloosed, and the gigantic machinery began to revolve.

This was a triumph worthy to be honored in the way that Englishmen love, by a little festivity; and as it chanced to be now ten years since Mr. Field had embarked in the enterprise, the pleasant thought occurred to him of getting his friends together to celebrate the anniversary. Accordingly, on the fifteenth of March, he invited them to dine together at the Palace Hotel. It was a pleasant occasion, calling forth the usual amount of toasts and speeches. Of the latter, those of Mr. Adams, the American Minister, and of John Bright, were widely copied in the United States. The next day was the annual meeting of the Atlantic Telegraph Company, when the Chairman, The Right Hon. James Stuart Wortley, thus referred to the gathering of the night before:

"Without saying any thing to detract from my deep source of gratitude to the other Directors, I cannot help especially alluding to Mr. Cyrus Field, who is present to-day, and who has crossed the Atlantic thirty-one times in the service of this Company, having celebrated at his table yesterday the anniversary of the tenth year of the day when he first left Boston in the service of the Company. Collected round his table last night was a company of distinguished men—members of Parliament, great capitalists, distinguished merchants and manufacturers, engineers and men of science, such as is rarely found together even in the highest house in this great metropolis. It was very agreeable to see an American citizen so surrounded. To me it was so personally, as it would have been to you, and it was still more gratifying, inasmuch as we were there to celebrate the approaching accomplishment of the Atlantic Telegraph."

We have no wish to repeat mere phrases of compliment; but it is always grateful to one who has toiled long and faithfully to carry out a public enterprise, to find in the hour of triumph that his labors are not forgotten. The Atlantic Telegraph had just passed through a critical period of its history. The enterprise had been in great danger of abandonment— at least for years. From this condition it had been rescued only by the most prompt and vigorous effort. How much this altered state of things was owing to the exertions of Mr. Field, let those speak who know best. At a meeting of the Directors of the Company, May fifth, 1864, on motion of Mr. C. M. Lampson, seconded by Mr. Francis Le Breton, it was unanimously

Resolved, "That the sincere thanks of this Board be given to Mr. Cyrus W. Field, for his untiring energy in promoting the general interests of the Atlantic Telegraph Company, and especially for his valuable and successful exertions during his present visit to Great Britain, in reference to the restoration of its financial position, and prospects of complete success."

Thus after infinite toil, the wreck of old disasters was cleared away, and the mighty task begun anew. The works of the Telegraph Construction and Maintenance Company are the largest in the world, and all their resources were now put in requisition. Never did greater care preside over a public enterprise. It was a case in which the motive of interest was seconded or overborne by pride and ambition. A cable was to be made to span the Atlantic Ocean, and to join the hemispheres; and it was a natural pride to produce a work that should be as nearly perfect as human skill could make it. The Scientific Committee, that had so long investigated the subject, had approved a particular form of cable, as "the one most calculated to insure success in the present state of our experimental knowledge respecting deep-sea cables," but at the same time recommended the utmost vigilance at every stage of the manufacture. These precautions deserve to be noted, as showing with what jealous care science watches over the birth of a great enterprise, and prescribes the conditions of success. They recommended:

That the conductivity of *the wire* should be fixed at a high standard, certainly not less than eighty-five per cent; that the cable should be at least equal to the best ever made; that the *core* should be electrically perfect; that it should be tested *under hydraulic pressure*, and at the highest pressure attainable in the tanks at the Company's works; that after this pressure, the core should be examined again, and before receiving its outer covering, be required to pass *the full electrical test under water;* that careful and frequent mechanical tests be made upon *the iron wire and hemp* as to their strength ; that special care be given to the *joints*, where different lengths of cable were spliced together ; and that when completed, the whole be tested under water for some length of time, at a temperature of seventy-five degrees. This was higher by forty degrees than the temperature of the Atlantic. The insulation is improved by cold ; so that, if it remained perfect in this warm water, it could not fail in the icy depths of the ocean.

After passing through such elaborate tests, all will be glad to see the final product of so much care and skill. As the long line begins to reel off from the great wheels and drums, we may examine it in its com-

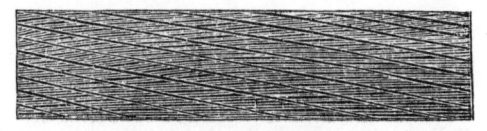

OLD ATLANTIC CABLE, 1858.

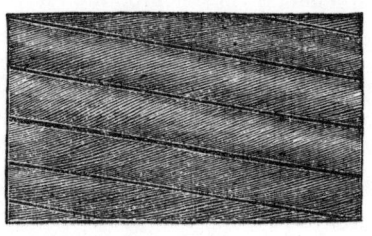

NEW ATLANTIC CABLE, 1865.

pleted and more perfect form. It is only necessary to compare it with the cable laid in 1858, to show its immense superiority. A glance at the two as they appear on the preceding page will show that the cable had *grown* since first it was planted in the ocean, as if it were a living product of the sea; or if we choose to consider it as a tendril of the British oak, the slender twig had become a tough, strong limb. This growth had been in every part, from core to circumference.

First, the central copper wire, which was the spinal cord, the nerve along which the lightning was to run, was nearly three times larger than before. The old conductor was a strand, consisting of seven fine wires, six laid round one, and weighed only one hundred and seven pounds to the mile. The new was composed of the same number of wires, but weighing three hundred pounds to the mile. This was made of the finest copper that could be obtained in the world—making a perfect conductor. To secure insulation, this conductor was first imbedded for solidity in Chatterton's compound, a preparation impervious to water, and then covered with four layers of gutta-percha, which were laid on alternately with four thin layers of Chatterton's compound. The old cable had but three coatings of gutta-percha, with nothing between. Its entire insulation weighed but two hundred and sixty-one pounds to the mile, while that of the new weighed four hundred pounds.

But a conductor ever so perfect, with insulation complete, was useless without proper external protection, to guard it against the dangers which must attend the long and difficult process of laying it across the ocean. The old cable had broken a number of times. The new must be made stronger. To this end it was incased with ten solid wires of the best iron, or rather, of a soft steel, like that used by Whitworth for his cannon. This made the cable much heavier than before. The old cable weighed but twenty cwt. to the mile, while the new one reached thirty-five cwt. and three quarters. But mere size and weight were nothing, except as they indicated increased strength. This was secured, not only by the larger iron wires, but by a further coating of rope. Each wire was surrounded separately with five strands of Manilla yarn, saturated with a preservative compound, and the whole laid spirally round the core, which latter was padded with ordinary hemp, saturated with the same preservative mixture. This rope covering was important in several respects. It kept the wires from coming in contact with the salt water, by which they might be corroded; and while it added greatly to the strength of the cable, it gave it also its own flexibility—so that while it had the strength of an iron chain, it had also the lightness and flexibility of a common ship's rope. This union of two qualities was all-important. The great problem had been to combine strength with flexibility. Mere dead weight was an objection. The new cable, though nearly twice as heavy as the old *in air*, when immersed in water, weighed but a trifle more; so that it was really much *lighter* in proportion to its size. This increased lightness was a very important matter in laying the cable, as it caused it to sink slowly. The old cable, though smaller, was heavy almost as a rod of iron, so that, as it ran out, it dropped at an angle which exposed it to great danger in case of a sudden lurch of the ship. Thus in 1857 it was broken by the stern of the Niagara being thrown up on a wave just as the brakes were shut down. Now the cable, being partially buoyed by the rope, would float out to a great distance from the ship, and sink down slowly in the deep waters.

By this combination of rope and iron, a cable was secured two and a half times as strong as the old—the breaking strain of the former having been three tons, five cwt., and of the latter seven tons and fifteen cwt. Or, to put it in another form, the contract strain of the former was less than five times its own weight per mile in water; so that if the cable had been laid in some parts of the Atlantic, where the ocean is more than five miles deep, it would have broken under the enormous strain. But the contract strain of the new cable was equal to *eleven* times its weight per mile in water, which, as the greatest depth of water to be passed was but two and a half miles, rendered the cable more than four times as strong as was required.

This great chain which was to bind the sea was to be twenty-three hundred nautical miles long, or nearly twenty-seven hundred statute miles! But where could this enormous bulk be stowed? Its weight would sink the Spanish Armada. In 1858, the cable loaded down two of the largest ships of war in the world, the Niagara and the Agamemnon. Yet now one much larger and bulkier was to be taken on board. This might have proved a serious embarrassment, but that Providence, which leads the progress of the race, prepares the means of advancement. It so happened that a few years before there had been built in England a ship of enormous proportions — the greatest floating thing constructed by the labor of man since Noah's ark was borne on the waters of the Deluge. The Great Eastern, whose iron walls had been reared by the genius of Brunel, had been for ten years waiting for "a mission." As a specimen of marine architecture she was perfect. She walked the waters in towering pride, scarce bending her imperial head to the waves that broke against her sides, as against the rocks of the shore. But with all her noble qualities, she was too great for the ordinary demands of commerce. Her very size was against her; and while smaller ships, on which she looked down with contempt, were continually flying to and fro across the sea, this leviathan,

> Hugest of all God's works
> That swim the ocean stream,

could find nothing worthy of her greatness. Here, then, was the vessel to receive the Atlantic cable.

Seeing her fitness for the purpose, a few of the gentlemen who were active in reviving the Atlantic Telegraph combined to purchase her, as she was about to be sold. One of them went down with all speed to Liverpool, and the next day telegraphed that the Big Ship was theirs. The new owners at once put her at the service of the Atlantic Company, with the express agreement that any compensation for her use should depend on the success of the expedition.

Next to the good fortune of finding such a ship ready to their hands, was that of finding an officer worthy to command her. Captain James Anderson, of the China, one of the Cunard steamers, had long been known to the travelling public, both of England and America, and no one ever crossed the sea with him without having awakened the strongest feeling of respect for his manly and seamanly qualities. A thorough master of his profession, having followed the sea for a quarter of a century, he was also a man of much general intelligence, and of no small scientific attainments. But it was something more than this which inspired such confidence. It was his ceaseless watchfulness. He always carried with him a feeling of religious responsibility for the lives of all on board, and for every interest committed to him. A man of few words, modest in manner, he was yet clear in judgment and prompt in action. This vigilance was especially marked in moments of danger. When a storm was gathering, all who saw that tall figure on the wheel-house, watching with a keen eye every spar in the ship and every cloud in the horizon, felt a new security from being under his care. Such was the man to be put in charge of a great expedition. He was the unanimous choice of the Board of Directors. The Cunard Company, with great generosity, consented to give up his services, valuable as they were, to forward an enterprise of such public interest. Being thus free, he accepted the trust, and entered upon it with enthusiasm. How well he fulfilled the expectations of all, the sequel will show.

The work now went on with speed. The wheels began to hum, and the great drums to reel off that line which, considering the distance it was to span, was hardly to be measured by miles, but rather by degrees of the earth's surface. Mere figures give but a vague impression of vast spaces. But it is a curious fact, ascertained by an exact computation, that if all the wires of copper and of iron, with the layers that made up the core and the outer covering, and the strands of yarn that were twisted into this one knotted sea-cable, were placed end to end, the whole length would reach from the earth to the moon!

As it came from the works in its completed state, it was plunged in water, to make it familiar with the element which was to be its future home. In the yards of the Company stood eight large tanks, which could hold each a hundred and forty miles. Here the cable was coiled to "hybernate," till it should be wanted for use the coming spring.

Seeing the work thus well under way, with no chance of another disastrous check, Mr. Field left England with heart at rest, and returned to America for the winter. But the first days of spring saw him again on the Atlantic. He reached England on the eighteenth of March. His visit was more satisfactory than a year before. The work was now well advanced. It was a goodly sight to go down to Morden Wharf at Greenwich, and see the huge machinery in motion, spinning off their leagues of deep-sea line. The triumph apparently was near at hand. It seemed indeed a predestined thing that the cable should finally be laid in the year of grace 1865—the end for which he had so faithfully toiled since 1858—seven weary years—as long as Jacob served for Rachel! But, less fortunate than Jacob, he was doomed to one more disappointment. At present, however, all looked well, and he could not but regard the prospect with satisfaction.

Having no more drudgery of raising money, he had now a few weeks' leisure to take a voyage up the Mediterranean. The canal across the Isthmus of Suez, which had been so long in progress, under the supervision of French engineers, was at length so far advanced that the waters of the Mediterranean were about to mingle with those of the Red Sea, and delegates were invited to be present from all parts of the world. An invitation had been sent to the Chamber of Commerce in New-York, and Mr. Field, then starting for Europe, was appointed as its representative. The visit was one of extraordinary interest. The occasion brought together a number of eminent engineers from every country of Europe, in company with whom this stranger from the New World visited the most ancient of kingdoms to see the spirit of modern enterprise invading the land of the Pyramids.

He returned to England about the first of May to find the work nearly completed. The cable was almost done, and a large part of it was already coiled on board the ship. This was an operation of much interest, which deserves to be described. The manufacture had begun on the first of September, and had gone on for eight months without ceasing, the works turning out fourteen miles a day even during the short days of winter. As the spring advanced, and the days grew longer, the amount was of course much increased. But by the last of January they had already accumulated about nine hundred miles of completed cable, when began the long and tedious work of transferring it to the Great Eastern. It was thus slow, because it could not be made directly from the yard to the ship. The depth of water at Greenwich was not such as to allow the Great Eastern to be brought up alongside the wharf. She was lying at Sheerness, thirty miles below, and the cable had to be put on board of lighters and taken down to where she lay in the stream. For this purpose the Admiralty had furnished to the Company two old hulks, the Iris and the Amethyst, which took their loads in turn. When the former had taken on board some two hundred and fifty tons of cable, she was towed down to the side of the Great Eastern, and the other took her place.

This was an operation which could not be done with speed. With all the men who could be employed, they coiled on board only about two miles an hour, or twenty miles a day—at which rate it would take about five months. The work began on the nineteenth of

January, early in the morning, and continued till June, before all was safely stowed on board. The Great Eastern herself had been fitted up to receive her enormous burden. It was an object to stow the cable in as few coils as possible. Yet it could not be all piled in one mass. Such a dead weight in the centre of the ship would cause her to roll fearfully. If coiled in one circle, it was computed that it would nearly fill Astley's theatre from the floor of the circus to the roof—making a pile fifty-eight feet wide and sixty feet high. To distribute this enormous bulk and weight, it was disposed in three tanks—one aft, one amidships, and one forward. The latter, from the shape of the ship, was a little smaller than the others, and held only six hundred and thirty-three miles of cable, while the two former held a little over eight hundred each. All were made of thick wrought-iron plates, and water-tight, so that the cable could be kept under water till it was immersed in the sea.

Thus with her spacious chambers prepared for the reception of her guest, the Great Eastern opened her doors to take in the Atlantic cable; and long as it was, and wide and high the space it filled, it found ample verge and room within her capacious sides. Indeed, it was the wonder of all who beheld it, how like a monster of the sea, she devoured all that other ships could bring. The Iris and the Amethyst came up time after time and disgorged their iron contents. Yet this Leviathan swallowed ship-load after ship-load, as if she could never be satisfied. A writer who visited her in May, when the cable was nearly all on board, was at a loss to find it. He looked along the deck, from stem to stern, but not a sign of it appeared. How he searched, and how the wonder grew, he tells in a letter to the "Railway News." After describing his approach to the ship, and climbing up her sides and his survey of her deck, he proceeds:

"But it is time that we should look after what we have mainly come to see, the telegraph cable. To our intense astonishment, we behold it nowhere, although informed that there are nearly two thousand miles of it already on board, and the remaining piece—a piece long enough to stretch from Land's End to John O'Groat's—is in course of shipment. We walk up and down on the deck of the Great Eastern without seeing this gigantic chain which is to bind together the Old and the New World; and it is only on having the place pointed out to us that we find where the cable lies and by what process it is taken on board. On the side opposite to where we landed, deep below the deck of our giant, there is moored a vessel surmounted by a timber structure resembling a house, and from this vessel the wonderful telegraph cable is drawn silently into the immense womb of the Great Eastern. The work is done so quietly and noiselessly, by means of a small steam-engine, that we scarcely notice it. Indeed, were it not pointed out to us, we would never think that that little iron cord, about an inch in diameter, which is sliding over a few rollers and through a wooden table, is a thing of world-wide fame—a thing which may influence the life of whole nations; nay, which may affect the march of civilization. Following the direction in which the iron rope goes, we now come to the most marvellous sight yet seen on board the Great Eastern. We find ourselves in a little wooden cabin, and look down, over a railing at the side, into an immense cavern below. This cavern is one of the three 'tanks' in which the two-thousand-mile cable is finding a temporary home. The passive agent of electricity comes creeping in here in a beautiful, silent manner, and is deposited in spiral coils, layer upon layer. It is almost dark at the immense depth below, and we can only dimly discern the human figures through whose hands the coil passes to its bed. Suddenly, however, the men begin singing. They intone a low, plaintive song of the sea; something like Kingsley's

'Three fishers went sailing away to the West,
Away to the West as the sun went down—'

the sounds of which rise up from the dark, deep cavern with startling effect, and produce an indescribable impression.

"We proceed on; but the song of the sailors who are taking charge of the Atlantic Telegraph cable is haunting us like a dream. In vain that our guide conducts us all over the big ship, through miles of galleries, passages, staircases, and promenades; through gorgeous saloons, full of mirrors, marbles, paintings, and upholstery, made 'regardless of expense;' and through buildings crowded with glittering steam apparatus of gigantic dimensions, where the latent power of coal and water creates the force which propels this monster vessel over the seas. In vain our attention is directed to all these sights; we do not admire them; our imagination is used up. The echo of the sailors' song in the womb of the Great Eastern will not be banished from our mind. It raises visions of the future of the mystic iron coil under our feet—how it will roll forth again from its narrow berth; how it will sink to the bottom of the Atlantic, or hang from mountain to mountain far below the stormy waves; and how two great nations, offsprings of one race and pioneers of civilization, will speak through this wonderful coil, annihilating distance and time. Who can help dreaming here, on the spot where we stand? For it is truly a marvellous romance of civilization, this Great Eastern and this Atlantic Telegraph cable. Even should our age produce nothing else, it alone would be the triumph of our age."

As the work approached completion, public interest revived in the stupendous undertaking, and crowds of wonder-seekers came down from London to see the preparations for the expedition. Even if not admitted on board, they found a satisfaction in sailing round the great ship, in whose mighty bosom was coiled this huge sea-serpent. It had also many distinguished visitors. Among others, the Prince of Wales came to see the ocean girdle which was to link the British islands with his future dominions beyond the sea.

At length, on the twenty-ninth of May, almost the last day of Spring, the manufacture of the cable was finished. The machines which for eight months had been in a constant whirl, made their last round. The tinkling of a bell announced that the machinery was empty, and the mighty work stood completed. It only remained that it should be got on board, and the ship prepared for her voyage. Hundreds of busy hands were at work without ceasing, and yet it was six weeks before she was ready to put to sea.

It may well be believed that it was no small affair to equip such an expedition. Beside the enormous burden of the cable itself, the Great Eastern had to take on board seven or eight thousand tons of coal, enough for a fleet, to feed her fires. Then she carried about five hundred men, for whom she had to make provision during the weeks they might be at sea. The stores laid in were enough for a small army. Standing on the wheel-house, and looking down, one might fancy himself in some large farm-yard of England. There stood the motherly cow that was to give them milk; and a dozen oxen, and twenty pigs, and a hundred and twenty sheep, while whole flocks of ducks and geese, and fowls of every kind, cackled as in a poultry-yard. Beside all this live stock, hundreds of barrels of provisions, of meats, and fruits, were stored in the well-stocked larder below. Thus laden for her voyage, the Great Eastern had in her a weight, including her own machinery, of twenty-one thousand tons—a burden almost as great as could have been carried by the whole fleet with which Nelson fought the battle of Trafalgar.

As the time of departure drew near, public curiosity was excited, and there was an extraordinary desire to witness the approaching attempt. The Company was besieged by applications from all quarters for permission to accompany the expedition. Had these requests been granted, on the scale asked, even the large dimensions of the Great Eastern could hardly have been sufficient for the crowds on board. The demand was most pressing for places for newspaper correspondents. These came not only from England, but from France and America. Almost every journal in London claimed the privilege of being represented. The result was what might have been expected. As it was impossible to satisfy *all*, and to discriminate in favor of some, and exclude others, would seem partial and unjust, they were finally obliged to exclude all. Of course this gave great offence. There was an outcry in England and in the United States at what was denounced as a selfish and suicidal policy. But it is doubtful whether any other possible course would have given better satisfaction.

Whether the Managers erred in this or not, it should be said that they applied the same inexorable rule to themselves—even directors of the Company being excluded, unless they had some special business on board.

It should be borne in mind that the expedition was not under the control of the Atlantic Telegraph Company at all, but of the Telegraph Construction and Maintenance Company, which had undertaken the work in fulfilment of a contract with the former Company to manufacture and lay down a cable across the Atlantic, in which it assumed the whole responsibility, not only making the cable, but chartering the ship and appointing the officers, and sending its own engineers to lay it down. Of course it had an enormous stake in the result. Hence it felt, not only authorized, but bound, to organize the expedition solely with reference to success. It was not a voyage of pleasure, but for business; for the accomplishment of a great and most difficult undertaking. Hence it was right that most strict rules should be adopted. Accordingly there was not a man on board who had not some business there. As the voyage promised to be one of the utmost practical interest to electricians and engineers, several young men were received as assistants in the testing-room or in the engineers' department; but there was no person who was not in some way engaged on the business of one or the other company, or connected with the management of the ship. Except Mr. Field, not an Atlantic Telegraph Director accompanied the expedition; and he represented also the Newfoundland Company. Mr. Gooch, M.P., was at once a Director of the Telegraph Construction and Maintenance Company, and Chairman of the Board that owned the Great Eastern, and so represented both those companies which had so great a stake in the result.

Thus the whole business was in the hands of the Telegraph Construction and Maintenance Company. It had its own officers to man the expedition— the captain and crew to sail the ship—its engineers to lay the cable—and its electricians to test it. Even the eminent electricians, Professor Thomson and Mr. Varley, who were on board in the service of the Atlantic Telegraph Company, were not allowed to interfere, *nor even to give advice* unless it were asked for in writing, and then it was to be given in writing. Their office was only to test the cable when laid, to pass messages through it from Newfoundland to Ireland, and to report it complete.

So rigorous were the rules which governed this memorable voyage. The whole enterprise was organized as completely as a naval expedition. Every man had his place. As when a ship is going into battle, every body is sent below that has not some business on deck, so it is not strange that in such a critical enterprise they did not want a host of supernumeraries on board.

Yet the Company was not unmindful of the anxiety of the public for news, and since it could not give a place to many correspondents, it engaged one, and that the best—W. H. Russell, LL.D., the well-known correspondent of the London Times, in the Crimea and in India. This brilliant writer was engaged to accompany the expedition—not to praise without discrimination, but to report events faithfully from day to day. He was accompanied by several artists, to illustrate the scenes of the voyage. Thus the Company made every provision to furnish information and even entertainment to the public. Several of these gentlemen afterward wrote accounts for different magazines—Blackwood, Cornhill, and Macmillan's. Their different reports, and especially the volume of Dr. Russell, which combines the accuracy and minuteness of a diary kept from day to day, with brilliant descriptions, set off by illustrations from drawings of the artists, furnish the public as full and complete an account as if there had been a special correspondent for every journal of England and America.

But if the public at large were very properly excluded, the organization on board was perfect and complete. At the head was Captain Anderson, of whom we have already spoken. As his duties would be manifold and increasing, he had requested the aid of an assistant commander, and Captain Moriarty, R. N., who had been in the Agamemnon in 1858, was permitted by the Admiralty to accompany the ship, and to give the invaluable aid of his experience and skill. The government also generously granted two ships of war, the Sphinx and the Terrible, to attend the Great Eastern. Thus the whole equipment of the expedition was English. Of the five hundred men on board the Great Eastern, there was but one American, and that was Mr. Field.

The engineering department was under charge of Mr. Samuel Canning, who, as the representative of the Telegraph Construction and Maintenance Company, was chief in command in all matters relating to laying the cable. For this responsible position no better man could have been chosen. Before the voyage was ended, he had ample opportunity to show his resources. He was ably seconded by Mr. Henry Clifford. Both these gentlemen had been on board the Agamemnon in the two Expeditions of 1858. They had since had large experience in laying submarine cables in the Mediterranean and other seas. It was chiefly by their united skill that the paying-out machinery had been brought to such perfection, that throughout the voyage it worked without a single hitch or jar. They had an invaluable helper in Mr. Temple.

The electrical department was under charge of Mr. De Sauty, who had had long experience in submarine telegraphs, and who was aided by an efficient corps of assistants. Professor Thomson and Mr. Varley, as we have said, represented the Atlantic Company. All these gentlemen had been unceasing in their tests of the cable in every form, both while in the process of manufacture and after it was coiled in the Great Eastern. The result of their repeated tests was to demonstrate that the cable was *many times more perfect than had been required*. With such marvellous delicacy did they test the current of electricity sent through it, that it was determined that of one thousand parts, over nine hundred and ninety-nine came out at the other end!

To complete this organization and equipment caused such delays as excited the impatience of all on board. But at length, when midsummer had fully come—at noon of Saturday, July fifteenth—the song of the sailors sounded the *chant du départ*. The Great Eastern was then lying at the Nore, and she seemed to cling to the English soil which she had griped with a huge Trotman weighing seven tons, held fast by a chain whereof every link weighed seventy pounds! To wrench this ponderous anchor from its bed required the united strength of near two hundred men. At last the bottom let go its hold, the anchor swings to the bow, the gun is fired, and the voyage is begun. A fleet of yachts and boats raise their cheers as the mighty hull begins to move. But mark how carefully she feels her way, following the lead of yonder little steamer, the Porcupine, the same faithful guide that seven years before led the Niagara up Trinity Bay one night when the faint light of stars twinkled on all the surrounding hills. Slowly they near the sea. Now the cliffs of Dover are in sight, and bidding her escort adieu, the Great Eastern glides along by the beautiful Isle of Wight, and then quickening her speed, with a royal sweep, she moves down the Channel. Off Falmouth she picked up the Caroline, a small steamer, which had left several days before with the shore end on board. She was laboring heavily with her burden, and made little headway in the rough waves. But the Great Eastern took her in tow, and she followed like a ship's boat in the wake of the monarch of the seas.

Thus they passed round to the coast of Ireland, to that Valentia Bay where, eight years before, the Earl of Carlisle gave his benediction on the departure of the Niagara and the Agamemnon, and where, a year later, the gallant English ship brought her end of the cable safely to the shore.

The point of landing had been changed from Valentia harbor five or six miles to Foilhommerum Bay, a wild spot where huge cliffs hang over the waves that here come rolling in from the Atlantic. On the top, an old tower of the time of Cromwell tells of the bloody days of England's great civil war. It is now but a mossy ruin. Here the peasants who flocked in from the country pitched their booths on the green sward, and looked down from the dizzy heights on the boats dancing in the bay below. At the foot of the cliff, a soft, sandy beach forms a bed for the cable, and here, as it issues from the sea, it is led up a channel which had been cut for it in the rocks.

As the shore end was very massive and unwieldy,

it could not be laid except in good weather; and as the sea was now rough, the Great Eastern withdrew to Bantry Bay, to be out of the way of the storms which sometimes break with fury on this rock-bound coast.

On Saturday this preliminary work was completed, the heavy shore end was carried from the deck of the Caroline across a bridge of boats to the beach, and hauled up the cliffs amid the shouts of the people. When once it was made fast to the rocks, the little steamer began to move and the huge coil slowly unwound, and like a giant awakened, stretched out its long iron arms. By half-past ten o'clock at night the hold was empty, the whole twenty-seven miles having been safely laid, and the end buoyed in seventy-five fathoms water. A dispatch was at once sent across the country to Bantry Bay to the Great Eastern to come around with all speed, and early the next morning her smoke was seen in the offing. Passing the harbor of Valentia, she proceeded to join the Caroline, which she reached about noon, and at once commenced splicing the massive shore end to her own deep-sea line. This was a work of several hours, so that it was toward evening before all was completed.

Thus, so many had been the delays of the past week, that it had come on to Sunday before the

Great Eastern was ready to begin her voyage. This—
which some might count a desecration of the holy
day—the sailors rather accepted as a good omen.
Had the shore end been laid forty-eight hours sooner,
the voyage might have begun on Friday, which sailors, who are proverbially superstitious, would have
thought an unlucky beginning. But Sunday, in their
esteem, is a good day. They like, when a ship is
moving out of sight of land, that the last sound from
the shore should be the blessed Sabbath bells. If that
sacred chime were not heard to-day, at least a Sabbath
peace rested on sea and sky. It was a calm summer's
evening. The sun was just sinking in the waves, as
the Great Eastern, with the two ships of war which
waited on either hand, to attend her royal progress,
turned their faces to the West, and caught the sudden
glory. Says Russell: "As the sun set, a broad stream
of golden light was thrown across the smooth billows
toward their bows, as if to indicate and illumine the
path marked out by the hand of heaven." What a
sacred omen! Had it been the fleet of Columbus sailing westward, every ship's company would have fallen
upon their knees on those decks, and burst forth in an
Ave Maria to the gentle Mistress of the Seas. We
trust in that manly crew there was many an eye that
took in the full beauty of the scene, and many a reverent heart that invoked a benediction.

In other respects the day was well chosen. It was the twenty-third of July. From the beginning, Captain Anderson had wished to sail on the twenty-third of June, or the twenty-second of July, so as to catch the full moon on the American coast. He desired also to take advantage of the westerly winds which prevail at that season, for in going against the wind the Great Eastern was steady as a rock. Every expectation was realized. To the Big Ship the ocean was as an inland lake. The paying-out machinery—the product of so much study and skill—worked "beautifully," and as the ship increased her speed, the cable glided into the water with such ease that it seemed but a holiday affair to carry it across to yonder continent. Such were the reflections of all that evening as the long summer twilight lingered on the sea. At midnight they went to sleep, to dream of an easy triumph.

Yet be not too confident. But a few hours had passed before the booming of a gun awoke all on board with the heavy tidings of disaster. The morning breaks early in those high latitudes, and by four o'clock all were on deck, with anxious looks inquiring for the cause of alarm. The ship was lying still, as if her voyage had already come to an end, and electricians, with troubled countenances, were passing in and out of the testing-room, which, as it was always kept darkened, looked like a sick-chamber where some royal patient lay trembling between life and death.

The method used by the electricians to discover a fault is one of such delicacy and beauty as shows the marvellous perfection of the instruments which science employs to learn the secrets of nature. The galvanometer is an invention of Professor Thomson, by which "a ray of light reflected from a tiny mirror suspended to a magnet travels along a scale, and indicates the resistance to the passage of the current through the cable by the deflection of the magnet, which is marked by the course of this speck of light. If the light of the mirror travels beyond the index, or out of bounds, an escape of the current is taking place, and what is technically called a fault has occurred." Such was the discovery on Monday morning. At a quarter past three o'clock the electrician on duty saw the light suddenly glide to the end of the scale and there vanish.

Fortunately it was not a fatal injury. It did not prevent signalling through the cable, and a message was at once sent back to the shore, giving notice of the check that had been received. But the electric current did not flow freely. There was a leak at some point of the line which it would not be prudent to pass over. They were now seventy-three miles from shore, having run out eighty-four miles of cable.

The tests of the electricians indicated the fault to be ten or a dozen miles from the stern of the ship. The only safe course was to go back and get this on board, and cut out the defective portion. It was a most ungrateful operation thus to be undoing their own work, but there was no help for it.

Such accidents had been anticipated, and before the Great Eastern left England, she had been provided with machinery to be used in case of necessity for "picking up" the cable. But this proved rather an unwieldy affair. It was at the bow, and as the paying-out machine was at the stern, the ship had to be got round, and the cable, which must first be cut, had to be transferred from one end to the other. This was not an easy matter. The Great Eastern was an eighth of a mile long, and to carry the cable along her sides for this distance, and over her high wheel-houses, was an operation at once tedious and difficult.

But at length the ship's head was brought round, and the end of the cable lifted over the bow, and grasped by the pulling-in machine, and the engine began to puff with the labor of raising the cable from the depths of the ocean. Fortunately they were only in four or five hundred fathoms water, so that the strain was not great. But the engine worked poorly, and the operation was very slow. With the best they could do, it was impossible to raise more than a mile an hour! But patience and courage, though it takes all day and all night!* The Great Eastern does her duty well, steaming slowly back toward Ireland, while the engine pulls, and the cable comes up, though reluctantly, from the sea, till on Tuesday morning at seven o'clock, when they have hauled in a little over ten miles, the cause of offence is brought on board. It is found to be a small piece of wire, not longer than a needle, that by some accident (for they did not then suspect a design) had been driven through the outer cover of the cable till it touched the core. There was the source of all the mischief. It was this pin's point which pricked this vital chord, opening a minute passage through which the electricity, like a jet of blood from a pierced artery, went streaming into the sea. It was with an almost angry feeling, as if to punish it for its intrusion, that this insignificant and contemptible source of trouble was snatched from its place, the wounded piece of cable was cut off, and a splice made and the work

* "All during the night the process of picking up was carefully carried on, the Big Ship behaving beautifully, and hanging lightly over the cable, as if fearful of breaking the slender cord which swayed up and down in the ocean. Indeed, so delicately did she answer her helm, and coil in the film of thread-like cable over her bows, that she put one in mind of an elephant taking up a straw in its proboscis."—RUSSELL.

of paying out renewed. But it was four o'clock in the afternoon of Tuesday before they were ready to resume the voyage. A full day and a half had been lost by this miserable piece of wire.

But the vexatious delay is over at last, and the stately ship, once more turning to the West, moves ahead with a steady composure, as if no petty trouble could vex her tranquil mind. Throughout the voyage the behavior of the ship was the admiration of all on board. While her consorts on either side were pitched about at the mercy of the waves, she moved forward with a grave demeanor, as if conscious of her mission, or as if eager to unburden her mighty heart, to throw overboard the great mystery that was coiled up within her, and to cast her burden on the sea.

The electricians, too, were elated, and with reason, at the perfection of the cable as demonstrated by every hour's experience. At intervals of thirty minutes, day and night, tests were passed from ship to shore, and to the delight of all, instead of finding the insulation weakened, it steadily improved as the cable was brought into contact with the cold depths of the Atlantic.

All now went well till Saturday, the twenty-ninth, when a little after noon there was again a cry from the ship, as if once more the cable were wounded and in pain. This time the fault was more serious than before. The electricians looked very grave, for they had struck "dead earth," that is, the insulation was completely destroyed, and the electric current was escaping into the sea.

As the fault had gone overboard, it was necessary to reverse their course, and haul in till the defective part was brought up from the bottom. This time it was more difficult, for they were in water two miles deep. Still the cable yielded slowly to the iron hands that drew it upward; and after working all the afternoon, about ten o'clock at night they got the fault on board. The wounded limb was at once amputated, and joining the parts that were whole, the cable was made new and strong again. Thus ended a day of anxiety. The next morning, which was the second Sabbath at sea, was welcomed with a grateful feeling after the suspense of the last twenty-four hours.

On Monday, the miles of cable that had been hauled up, and which were lying in huge piles upon the deck, were subjected to a rigid examination, to find out where the fault lay. This was soon apparent. Near the end was found a piece of wire thrust through its very heart, as if it had been driven into it. All looked black when this was discovered, for at once it excited suspicions of design. It was remarked that the same gang of workmen were in the tank as at the time of the first fault. Mr. Canning sent for the men, and showing them the cable pierced through with the wire, asked them how it occurred. very man replied that *it must have been done by design*, even though they accused themselves, as this implied that there was a traitor among them. It seemed hard to believe that any one could be guilty of such devilish malignity. Yet such a thing had been done before in a cable laid in the North Sea, where the insulation was destroyed by a nail driven into it. The man was afterward arrested, and confessed that he had been hired to do it by a rival company. The matter was the subject of a long investigation in the English courts. In the present case there were many motives which might prompt to such an act. The fall in the stock on the London Exchange, caused by a loss of the cable, could hardly be less than half a million sterling. Here was a temptation such as betrays bold, bad men into crime. However, as it was impossible to fix the deed on any one, nothing was proved, and there only remained a painful suspicion of treachery. Against this it was their duty to guard. Therefore it was agreed that the gentlemen on board should take turns in keeping watch in the tank. It was very unpleasant to Mr. Canning thus to set a watch on men, many of whom had been with him in his former cable-laying expeditions, but the best of them admitted the necessity of it, and were as eager as himself to find out the Judas among them.

But accident or villainy, it was defeated this time, and the Great Eastern proudly continued her voyage. Not the slightest check interrupted their progress for the next three days, during which they passed over five hundred miles of ocean. It was now they enjoyed their greatest triumph. They were in the middle of the Atlantic, and thus far the voyage had been a complete success. The ship seemed as if made by Heaven to accomplish this great work of civilization. The paying-out apparatus was a piece of mechanism to excite the enthusiasm of an engineer, so smoothly did its well-oiled wheels run. The strain never exceeded fourteen hundred-weight, even in the greatest depths of the Atlantic. And as for the cable itself, it seemed to come as near perfection as it was possible to attain. As before, the insulation was greatly improved by submergence in the ocean. With every lengthening league it grew better and better. It seems almost beyond belief, yet the fact is fully attested that, when in the middle of the ocean, the communication was so perfect that they could tell at Valentia every time the Great Eastern rolled.* With such omens of suc-

* So exquisitely sensitive was the copper strand, that as the Great Eastern rolled, and so made the cable pass across the magnetic meridian, the induced current of electricity, incomprehensibly faint as it

318 HISTORY OF THE ATLANTIC TELEGRAPH.

cess, who could but feel confident? And when on Monday they passed over a deep valley, where lay "the bones of three Atlantic cables," it was with a proud assurance that they should not add another to the number.

But Wednesday brought a sudden termination of their hopes. They had run out about twelve hundred miles of cable, and were now within six hundred miles of Newfoundland. Two days more would have made them safe, as it would have brought them into the shallow waters of the coast. Thus it was when least expected that disaster came. We shall give very briefly the record of this fatal day. In the morning, while Mr. Field was keeping watch in the tank, with the same gang of men who had been there when the trouble occurred before, a grating sound was heard, as if a piece of wire had caught in the machinery, and word was passed up to the deck to look out for it; but the caution seems not to have been heard, and it passed over the stern of the ship. Soon after a report came from the testing-room of "another fault." It was not a bad one, since it did not prevent communication with land; and much anxiety might have been saved had a message been sent to Ireland that they

must have been, produced nevertheless a perceptible deviation of the ray of light on the mirror galvanometer at Foilhommerum.—*London Times.*

were about to cut the cable, in order to haul it on board. But small as the fault was, it could not be left behind. Down on the deep sea-floor was some minute defect, a pin's point in a length measured by thousands of miles. Yet that was enough. Of this marvellous product of human skill, it might in truth be said, that it was like the law of God in demanding absolute perfection. To offend in one point was to be guilty of all.

This new fault, though it was annoying, did not create alarm, for they had been accustomed to such things, and regarded them only as the natural incidents of the voyage. Had the apparatus for pulling in been complete, it could not have delayed them more than a few hours. But this had been the weak point of the arrangements from the beginning—the *bête noire* of the expedition. The only motive power was a little donkey engine, (rightly named,) which puffed and wheezed as if it had the asthma. This was now put in requisition, but soon gave out for want of more steam. While waiting for this a breeze sprang up, which caused the Great Eastern to drift over the cable, by which it was badly chafed, so that when it was hauled in, as the injured part was coming over the bows and was almost within grasp, suddenly *it broke* and plunged into the sea!

Thus it came without a moment's warning. So unexpected was such a catastrophe, that the gentlemen had gone down to lunch, as it was a little past the hour of noon. But Mr. Canning and Mr. Field stood watching the cable as it was straining upward from the sea, and saw the snapping of that cord, which broke so many hopes. The impression may be better imagined than described. Says a writer on board: "Suddenly Mr. Canning appeared in the saloon, and in a manner which caused every one to start in his seat, said, 'It is all over! It is gone!' then hastened onward to his cabin. Ere the thrill of surprise and pain occasioned by these words had passed away, Mr. Field came from the companion into the saloon, and said, with composure admirable under the circumstances, though his lip quivered and his cheek was blanched, 'The cable has parted and has gone overboard.' All were on deck in a moment, and there, indeed, a glance revealed the truth."

At last it had come—the calamity which all had feared, yet that seemed so far away only a few hours before. Yet there it was—the ragged end on board, torn and bleeding, the other lying far down in its ocean grave.

Here in America, of course, nothing could be known of the fate of the expedition till its arrival on our shores. But in England its progress was reported from day to day, and as the success up to this point had raised the hopes of all to the highest pitch, the sudden loss of communication with the ship was a heavy blow to public expectation, and gave rise to all sorts of conjectures. At first a favorite theory was, that communication had been interrupted by a magnetic storm. These are among the most mysterious phenomena of nature—so subtle and fleeting as to be almost beyond the reach of science. No visible sign do they give of their presence. No clouds darken the heavens; no thunder peals along the sky. Yet strange influences trouble the air. At this very hour, Professor Airy, the Astronomer Royal at the Observatory at Greenwich, reported a magnetic storm of unusual violence. Said a London paper: "Just when the signals from the Great Eastern ceased, a magnetic storm of singular violence had set in. Unperceived by us, not to be seen in the heavens, nor felt in the atmosphere, the earth's electricity underwent a mysterious disturbance. The recording instruments scattered about the kingdom, everywhere testified to the fury of this voiceless tempest, and there is every reason to suppose that the confusion of signals at midday on Wednesday was due to the strange and unusual earth-currents of magnetism, sweeping wildly across the cable as it lay in apparently untroubled waters at the bottom of the Atlantic."

Said the Times: "At Valentia, on Wednesday last,

the signals, up to nine A.M., were coming with wonderful distinctness and regularity, but about that time a violent magnetic storm set in. No insulation of a submarine cable is ever so perfect as to withstand the influence of these electrical phenomena, which correspond in some particulars to storms in the ordinary atmosphere, their direction generally being from east to west. Their action is immediately communicated to all conductors of electricity, and a struggle set up between the natural current and that used artificially in sending messages. This magnetic storm affected every telegraphic station in the kingdom. At some the wires were utterly useless; and between Valentia and Killarney the natural current toward the west was so strong along the land lines that it required an addition of five times the ordinary battery power to overcome it. This magnetic storm, which ceased at two A.M. on Friday, was instantly perceptible in the Atlantic cable."

But these explanations, so consoling to anxious friends on land, did not comfort those on board the Great Eastern. They knew, alas! that the cable was at the bottom of the ocean, and the only question was, if any thing could be done to recover it.

Now began a work of which there had been no example in the annals of the sea. The intrepid Canning declared his purpose to grapple for the cable! The proposal seemed wild, dictated by the frenzy of despair. Yet he had fished in deep waters before. He had laid his hand on the bottom of the Mediterranean, but that was a shallow lake compared with the depths into which the Atlantic cable had descended. The ocean is here two and a half miles deep. It was as if an Alpine hunter stood on the summit of Mont Blanc and cast a line into the vale of Chamouni. Yet who shall put bounds to human courage? The expedition was not to be abandoned without a trial of this forlorn hope. There were on board some five miles of wire rope, intended to hold the cable in case it became necessary to cut it and lash it to the buoys, to save it from being lost in a storm. This was brought on deck for another purpose. "And now came forth the grapnels, two five-armed anchors, with flukes sharply curved and tapered to a tooth-like end—the hooks with which the Giant Despair was going to fish from the Great Eastern for a take worth, with all its belongings, more than a million." These huge grappling-irons were firmly shackled to the end of the rope, and brought to the bows and thrown overboard. One splash, and the whole has disappeared in the bosom of the ocean. Down it goes—deeper, deeper, deeper still. For two full hours it continued sinking before it struck the earth, and like a pearl-diver, began searching for its lost treasure on the bottom of the sea.

What did it find there? The wrecks of ships that had gone down a hundred years ago, with dead men's bones whitening in the deep sea caves? It sought for something more precious to the interest of civilization than gems and gold.

The ship was now a dozen miles or so from the place of accident. The cable had broken a little after noon, when the sun was shining clear, so that Captains Anderson and Moriarty had just obtained a perfect observation, from which they could tell, within half a mile, the very spot where it had gone down. To reach it now, with any chance of bringing it up, it would be necessary to hook it a few miles from the end. It had been paid out in a line from east to west. To strike it broadside, the ship stood off in the afternoon a few miles to the south. Here the grapnel was thrown over about three o'clock, and struck bottom about five, when the ship began slowly drifting back on her course. All night long those iron fingers were raking the bottom of the deep but grasping nothing, till toward morning the long rope quivered like a fisherman's line when something has seized the end, and the head of the Great Eastern began to sway from her course, as if it felt some unseen attraction. As they begin to haul in, the rapidly increasing strain soon renders it certain that they have got hold of *something*. But what can it be? How do they know it is their lost cable? This question has often been asked since. They did not *see* it. How do they know that it was not the skeleton of a whale, or a mast or spar, the fragment of a wrecked ship? This question is easily answered. If it had been any loose object which was being drawn up from the sea, its weight would have diminished as it came nearer the surface. But on the contrary, the strain, as shown by the dynamometer, steadily *increased*. This could only be from some object lying prone on the bottom. To an engineer the proof is like a mathematical demonstration.

Having then caught the cable, they had good hopes of getting it again, their confidence increasing with every hundred fathoms brought on board. For hours the work went on. They had raised it seven hundred fathoms—or three quarters of a mile—from the bottom, when an iron swivel gave way, and the cable once more fell back into the sea, carrying with it nearly two miles of rope.

The first attempt had failed, but the fact that they had unmistakably caught the prize gave them courage for a second. Preparations were at once begun, but fogs came on and delayed the attempt till Monday, when it was repeated. This time the grapnel caught again. It was late in the afternoon when it got its hold, and the work of pulling in was kept up all night. But as the sea was calm and the moon shining brightly, all joined in it with spirit, feeling elated with the hope of triumph on the morrow.

That was not to be; but each attempt seemed to come nearer and nearer to victory. This time the cable was drawn up a full mile from the bottom, and hung suspended a mile and a half below the ship. Had the rope been strong enough, it might have been brought on board. But again a swivel gave way, and the cable, whose sleep had been a second time disturbed, sought its ocean bed.

These experiments were fast using up the wire rope, and every expedient had to be resorted to, to piece it out and to give it strength. Each shackle and swivel was replaced by new bolts, and the capstan was increased four feet in diameter, by being belted with enormous plates of iron, to wind the rope around it, if the picking up machinery should fail. This gave full work to all the mechanics on board. The ship was turned into a very cave of Vulcan, presenting at night a scene which might well take the eye of an artist, and which Russell thus describes :

" The forge fires glared on her decks, and there, out in the midst of the Atlantic, anvils rang and sparks flew; and the spectator thought of some village far away, where the blacksmith worked, unvexed by cable anxieties and greed of speedy news. As the blaze shot up, ruddy, mellow and strong, and flung arms of light aloft and along the glistening decks, and then died into a red centre, masts, spars, and ropes were for the instant touched with a golden gleaming, and strange figures and faces were called out from the darkness—vanished, glinted out again—rushed suddenly into foreground of bright pictures, which faded soon away—flickered—went out—as they were called to life by its warm breath, or were buried in the outer darkness! Outside all was obscurity, but now and then vast shadows, which moved across the arc of the lighted fog-bank, were projected far away by the flare; and one might well pardon the passing mariner, whose bark drifted him in the night across the track of the great ship, if, crossing himself, and praying with shuddering lips, he fancied he beheld a phantom ship freighted with an evil crew, and ever after told how he had seen the workshops of the Inferno floating on the bosom of the ocean."

While preparing for a third attempt, the ship had been drifting about, sometimes to a distance of thirty or forty miles, but it had marked the course where the cable lay by two buoys thrown over about ten miles apart, each bearing a flag which might be seen at a distance, and so easily came back to the spot. On Thursday morning all was ready, and the line was cast as before, but after some hours of drifting, it was evident that the ship had passed over the cable without grappling. The line was hauled in, and the reason at once appeared. One of the flukes had caught in the chain, so that it could not strike its teeth into the bottom. This was cleared away, and the rope prepared for a fourth and final attempt.

It was at noon of Friday that the grapnel went overboard for the last time. By four o'clock it had caught, and the work of hauling in recommenced. Again the cable was brought up nearly eight hundred fathoms, when the rope broke, carrying down two miles of its own length, and with it the hopes of the Atlantic Telegraph for the present year.

Their resources were exhausted. For nine days had that heroic crew persevered in their attempt. Or rather (for they scarce took note of day or night in the excitement of that long struggle) we might say,

> Nine times the space
> That measures day and night to mortal men,

they kept up the contest. There is something grand though sad in the spectacle of this noble ship thus lingering in mid-ocean, moving round and round one fatal spot, with her great eye fixed on the place where her treasure is gone down, and vainly striving to wrest it from the hand of the spoiler.

Baffled they were, yet they had not toiled in vain. They had shown what the power of man can do, that henceforth it was not to be bounded by any thing on the solid earth, or in the waters under the earth. Three times they had grasped the prize, and each time failed to recover it, only for want of ropes strong enough to bring it on board. *The cable never broke.* Herein it proved its strength, and gave good omen for future success.

But for the present all was over. The attempt must be abandoned for the year 1865, *but not for ever;* and with this purpose in her "constant mind," the Great Eastern swung sullenly around, and turned her imperial head toward England, like a warrior retiring from the field—not victorious, nor yet defeated and despairing, but with her battle-flag still flying, and resolved once more to attempt the conquest of the sea.

CHAPTER XV.

RESULT OF THE EXPEDITION OF 1865. CONFIDENCE STRONGER THAN EVER. INSTANT RESOLVE TO RAISE THE BROKEN END OF THE CABLE, TO COMPLETE IT TO NEWFOUNDLAND, AND TO LAY ANOTHER LINE. NEW SHARES ISSUED. METHOD DECLARED UNAUTHORIZED BY LAW. FORMATION OF THE ANGLO-AMERICAN TELEGRAPH COMPANY. CAPITAL RAISED. NEW CABLE MADE AND SHIPPED ON BOARD THE GREAT EASTERN.

THE expedition of 1865, though not an immediate success, yet had the moral effect of a victory, as it confirmed the most sanguine expectations of all who embarked in it. The great experiment made during those four weeks at sea, had demonstrated many points which were most important elements in the problem of Ocean Telegraphy. These are summed up in the following paper, which was signed by persons officially engaged on board the Great Eastern:

1. It was proved by the expedition of 1858, that a Submarine Telegraph Cable could be laid between Ireland and Newfoundland, and messages transmitted through the same.

By the expedition of 1865 it has been fully demonstrated:

2. That the insulation of a cable improves very much after its submersion in the cold deep water of the Atlantic, and that its conducting power is considerably increased thereby.

3. That the steamship Great Eastern, from her size and constant steadiness, and from the control over her afforded by the joint use of paddles and screw, renders it safe to lay an Atlantic Cable in any weather.

4. That in a depth of over two miles four attempts were made to grapple the cable. In three of them the cable was caught by the grapnel, and in the other the grapnel was fouled by the chain attached to it.

5. That the paying-out machinery used on board the Great Eastern worked perfectly, and can be confidently relied on for laying cables across the Atlantic.

6. That with the improved Telegraphic instruments for long submarine lines, a speed of more than eight words per minute can be obtained through such a cable as the present Atlantic between Ireland and Newfoundland, as the amount of slack actually paid out did not exceed fourteen per cent, which would have made the total cable laid between Valentia and Heart's Content nineteen hundred miles.

7. That the present Atlantic Cable, though capable

of bearing a strain of seven tons, did not experience more than fourteen hundred-weight in being paid out into the deepest water of the Atlantic between Ireland and Newfoundland.

8. That there is no difficulty in mooring buoys in the deep water of the Atlantic between Ireland and Newfoundland, and that two buoys even when moored by a piece of the Atlantic Cable itself, which had been previously lifted from the bottom, have ridden out a gale.

9. That more than four nautical miles of the Atlantic Cable have been recovered from a depth of over two miles, and that the insulation of the gutta-percha covered wire was in no way whatever impaired by the depth of water or the strains to which it had been subjected by lifting and passing through the hauling-in apparatus.

10. That the cable of 1865, owing to the improvements introduced into the manufacture of the gutta-percha core, was more than one hundred times better insulated than cables made in 1858, then considered perfect and still working.

11. That the electrical testing can be conducted with such unerring accuracy as to enable the electri-

cians to discover the existence of a fault immediately after its production or development, and very quickly to ascertain its position in the cable.

12. That with a steam-engine attached to the paying-out machinery, should a fault be discovered on board, whilst laying the cable, it is possible that it might be recovered before it had reached the bottom of the Atlantic, and repaired at once.

    S. CANNING, Engineer-in-Chief, Telegraph Construction and Maintenance Company.
    JAMES ANDERSON, Commander of the Great Eastern.
    HENRY A. MORIARTY, Staff Commander, R. N.
    DANIEL GOOCH, M.P., Chairman of " Great Ship Co."
    HENRY CLIFFORD, Engineer.
    WILLIAM THOMSON, LL.D., F.R.S., Prof. of Natural Philosophy in the University of Glasgow.
    CROMWELL F. VARLEY, Consulting Electrician Electric and International Telegraph Co.
    WILLOUGHBY SMITH.
    JULES DESPECHER.

This was a grand result to be attained in one short month; and if not quite so gratifying as to have the cable laid at once, and the wire in full operation, yet as it settled the chief elements of success, we say that in its moral influence it had the inspiring effect of a victory. All who were on that voyage felt a confidence such as they had never felt before. They came back, not desponding and discouraged, but buoyant with hope, and ready at once to renew the attempt.

This confidence appeared at the first meeting of directors. The feeling was very different from that after the return of the first expedition of 1858. So animated were they with hope, and so *sure* of success the next time, that all felt that one cable was not enough, they must have *two*, and so it was decided to take measures not only to raise the broken end of the cable and to complete it to Newfoundland, but also to construct and lay an entirely new one, so as to have a double line in operation the following summer.

The contractors, partaking the general confidence, came forward promptly with a new offer even more liberal than that made before. They proposed to construct a new line, and to lay it across the Atlantic for half a million sterling, which was estimated to be the actual cost to them, reserving all compensation to themselves to depend on success. If successful, they were to receive twenty per cent on the cost, or one hundred thousand pounds, to be paid in shares of the Company. They would engage, also, to go to sea fully prepared to raise the broken end of the cable now in mid-ocean, and with a sufficient length, including that on board the Great Eastern, to complete the line to Newfoundland. Thus the Company would have two cables instead of one.

In this offer the contractors assumed a very large risk. They now went a step further, and in the contingency of the capital not being raised otherwise, they offered *to take it all themselves*—to lay the line at their own risk, and to be paid only in the stock of the Company, which, of course, must depend for its value on the success of the next expedition. It was finally resolved to raise six hundred thousand pounds of new capital by the issue of a hundred and twenty thousand shares of five pounds each, which should be preferential shares, entitled to a dividend of twelve per cent before the eight per cent dividend to be paid on the former preference shares, and the four per cent on the ordinary stock. This was offering a substantial inducement to the public to take part in the enterprise, and it was thought with reason that this fresh issue of stock, though it increased the capital of the Company, yet as it was all to be employed in forwarding the great work, would not only create new property, but give value to the old. The proposal of the manufacturers was therefore at once accepted by the Directors, and the work was instantly begun. Thus hopeful was the state of affairs when Mr. Field returned to America in September.

But he was never easy to be long out of sight of his beloved cable, and so three months after he went back to England, reaching London on the twenty-fourth of December. He came at just the right moment, for the Atlantic Telegraph was once more in extremity Only two days before the Attorney-General of England had given a written opinion that the Company *had no legal right* to issue new twelve per cent preference shares, and that such issue could only be authorized by an express act of Parliament. This was a fatal decree to the Company. It was the more unexpected, as, before offering the twelve per cent capital, they had been fortified by the opinions of several eminent lawyers and solicitors in favor of the legality of their proceedings. It invalidated not only what they were going to do, but what they had done already. Hence, as the effect of this decision, all the works were stopped, and the money which had been paid in was returned to the subscribers.

This was a new dilemma, out of which it was not easy to find a way of relief. Parliament was not in session, Lords and Commons being away in the country keeping the Christmas holidays. Even if it had been, the time for applying to it had passed, as a notice of any private bill to be introduced must be given before the thirtieth of November, which was gone a month ago. To wait for an act of Parliament, therefore, would inevitably postpone the laying of the cable for another year. So disheartening was the prospect at the close of 1865.

But they had seen dark days before, and were not to give it up without a new effort. Happily, the cause had strong friends to stand by it even in this crisis of suspended animation. One of these to whom Mr. Field now went for counsel, was Mr. Daniel Gooch, M.P., a gentleman well known in London, as one of that class of engineers formed in the school of Stephenson and Brunel, who have risen to the position of great capitalists, and who, by their enterprise and wealth, have done so much to develop the resources of England. He was Chairman of the Great Western Railway. His connection with the Atlantic Telegraph was somewhat curious. Until lately he had not full confidence in its success. Though a man of large fortune, and a personal friend of Mr. Field, the latter had never prevailed on him to subscribe a single pound. But he went out on the expedition of '65, as chairman of the company that owned the Great Eastern; and what he then saw convinced him. He came back fully satisfied; he knew it could be done, and was ready to prove his faith by his works. Consulting on the present difficulty, he suggested that the only relief was *to organize a new Company*, which should assume the work, and which could issue its own shares and raise its own capital. This opinion was confirmed by that eminent legal authority, Mr. John Horatio Lloyd. To such a Company Mr. Gooch said he would subscribe £20,000; Mr. Field put down £10,000.

Next, he betook himself to that prince of English capitalists, Mr. Thomas Brassey, who heard from his lips for the first time, that the affairs of the Atlantic Telegraph Company had suddenly come to a standstill. At this he was much surprised, but instantly cheered his informer by saying: "Mr. Field, don't be discouraged; go down to the Company, and tell them to go ahead, and whatever the cost, I will bear one tenth of the whole." Who *could* be discouraged with such a Richard the Lion-hearted to cheer him on?

Meetings were called of the Directors of both the Atlantic Company and the Telegraph Construction and Maintenance Company; and frequent conferences were held between them. The result was the formation of a new company called the ANGLO-AMERICAN TELEGRAPH COMPANY, with a capital of £600,000, which contracted with the Atlantic Company to manufacture and lay down a cable in the summer of 1866, for doing which it is to be entitled to what virtually amounts to a preference dividend of twenty-five per cent: as a first claim is secured to them by the Atlantic Telegraph Company upon the revenue of the cable or cables (after the working expenses have been provided for) to the extent of £125,000 per annum; and the New-York, Newfoundland, and London Telegraph Company undertake to contribute from their revenue a further annual sum of £25,000, on condition that a cable shall be at work during 1866; an agreement to this effect having been signed by Mr. Field, subject to ratification by the Company in New-York, which was obtained as soon as the steamer could cross the ocean and bring back the reply.

The terms being settled, it remained only to raise the capital. The Telegraph Construction and Maintenance Company led off with a subscription of £100,000. This was followed by the names of ten gentlemen, who put down £10,000 apiece. Of these Mr. Gooch declared his willingness to increase his to £20,000, and Mr. Brassey to £60,000, if it were needed. Mr. Henry Bewley, of Dublin also, who was already a very large owner of the Atlantic stock, declared his readiness to put down £20,000 more. But it was not found to be necessary. And so they all stood at £10,000. The names of these ten subscribers deserve to be given, as showing who stood forward to save the cause in this crisis of its fate. They were as follows:

Henry Ford Barclay, Henry Bewley, Thomas Brassey, A. H. Campbell, M.P., George Elliot, Cyrus W. Field, Richard Atwood Glass, Daniel Gooch, M.P., John Pender, M.P., and John Smith, Esqs. There were four subscriptions of £5000, namely, Thomas

Bolton and Sons, James Horsfall, Esq., A Friend of Mr. Daniel Gooch, M.P., and John and Edwin Wright; one of £2500 by John Wilkes and Sons; three of £2000 by C. M. Lampson, J. Morison, and Ebenezer Pike, Esqs.; and two of £1000 by Edward Cropper, and Joseph Robinson, Esqs.—making in all £230,500.

These were all private subscriptions made before even the prospectus was issued, or the books opened to the public. After such a manifestation of confidence, we are not surprised to learn that the whole of the capital required to proceed with the cable was secured within fourteen days. This was a great triumph to be obtained, especially at a time of general depression in commercial affairs in England.

And now once more the work begins. No time is to be lost. It is already the first of March, and but four months remain to manufacture sixteen hundred and sixty nautical miles of cable, and to prepare for sea. But the obstacles once cleared away, all sprang to their work with new hope and vigor.

In the cable to be made for the new line, there was but little change from that of the last year, which had proved nearly perfect. Science, however, aided by experience, was constantly devising some improvement. So now, while the general form and size were retained, a slight change in the outer covering was found to make the cable both lighter and stronger.

The iron wires were *galvanized*, which secured them perfectly from rust or corrosion by salt water. Thus protected, they could dispense with the preservative mixture of the former year. This left the cable much cleaner and whiter. Instead of its black coat, it had the fresh, bright appearance of new rope. It had another advantage. As the tarry coating was sticky, slight fragments of wire might adhere to it, and do injury, a danger to which the new cable was not exposed. At the same time, galvanizing the wires gave them greater ductility, so that in the case of a heavy strain the cable would *stretch* longer without breaking. By this alteration it was rendered more than four hundred-weight lighter per mile, and would bear a strain of nearly half a ton more than the one laid the year before.

The machinery also was perfected in every part, to withstand the great strain which might be brought upon it in grappling and lifting the cable from the great depths of the Atlantic. This necessitated almost a reconstruction of the machinery, together with engines of greater power, applied both to the gear for hauling in forward and that for paying out aft. Thus, in case of a fault, the motion of the ship could be easily reversed, and the cable hauled back by the paying-out machinery, without waiting for the long and tedious process of bringing the cable round from the stern to the bow of the ship.

But the most marvellous improvement had been in the method of testing the cable for the discovery of faults. In the last expedition, a grave omission had been in the long intervals during which the cable was left without a test of its insulation. Thus, from thirty to thirty-five minutes in each hour it was occupied with tests of minor importance, which would not indicate the existence of a fault. Hence, if a fault occurred on ship-board, it might pass over the stern, and be miles away before it was discovered. But now a new and ingenious method was devised by Mr. Willoughby Smith, by which the cable will be tested *every instant.* The current will not cease to flow any more than the blood ceases to flow in human veins. The chord is vital in every part, and if touched at any point it reveals the wound as instinctively as the nerves of a living man flash to the brain a wound in any part of the human frame.

The process of detecting faults is too scientific to be detailed in these pages. We can only stand in silent wonder at the result, when we hear it stated by Mr. Varley, that the system of testing is brought to such a degree of perfection, that skilful electricians can point out minute faults with an unerring accuracy, "even when they are so small that they would not weaken the signals through the Atlantic cable one millionth part!"

Another marvellous result of science was the exact report obtained of the state of that portion of the cable now lying in the sea. The electricians at Valentia were daily experimenting on the line which lay stretched twelve hundred and thirteen miles on the bottom of the deep, and pronounced it intact. Not a fault could be found from one end to the other. As when a master of the organ runs his hands over the keys, and tells in an instant if it be in perfect tune, so did these skilful manipulators, fingering at the end of this mightier instrument, declare it to be in perfect tone, ready to whisper its harmonies through the seas. At the same time. the ten hundred and seventy miles of cable left on board the Great Eastern were pronounced as faultless as the day it had been shipped on board.

With such conclusions of science to animate and inspire them, the great task of manufacturing nearly seventeen hundred miles of cable once more began. And while this work went on, the Great Eastern, that had done her part so well before, again opened her sides, and the mysterious cord was drawn into her vast, dark, silent womb, from which it was to issue only into the darker and more silent bosom of the deep.

CHAPTER XVI.

The Expedition of 1866. Victory at Last.

In these pages we have led our readers through twelve long years, and have had to tell many a tale of disaster and defeat. It is now our privilege to tell of triumphant success. Victory has come at last, but not by the chance of fortune, but by the utmost efforts of man, by the union of science and skill with indomitable perseverance. The failure of the last year was a sad disappointment; but so far from damping the courage of those embarked in the enterprise, it only roused them to a more gigantic effort. They were now to prepare for a fifth expedition. In this they set themselves to anticipate every possible emergency, and to combine the elements of success so as to render failure impossible.

The Great Eastern herself, which they had come to regard with a kind of fondness, a feeling of affection and pride, as the ark that was to bear their fortunes across the deep, was made ready for her crowning achievement. For months Captain Anderson and Mr. Halpin, his chief officer, worked day and night to get her into perfect trim. She had become sadly fouled in her many voyages. As she swam the seas, a thousand things clung to her as to a floating island, till her hull was encrusted with mussels and barnacles two feet thick, and long seaweed flaunted from her sides. Like a brave old war-horse, long neglected, she needed a thorough grooming, to have her hair combed and her limbs well rubbed down, to fit her to take the field. But it was not an easy matter to get under the huge creature, to give her such a dressing. Yet Captain Anderson was equal to the emergency. He contrived a simple instrument by which every part of her bottom was raked and scrubbed. Getting rid of this rough, shapeless mass would make her feel easy and comfortable at sea, and add at least a knot an hour to her speed.

The boilers too were thoroughly cleansed and repaired in every part, and the paddle-engines were so arranged that in five minutes they could be disconnected, so that by going ahead with one and backing with the other, the ship could be held perfectly at rest or be turned around in her own length, a very important matter when they should come to fish in deep waters for the broken end of the cable. To prepare for this, she was armed with chains and ropes and irons of the most formidable kind. For grappling the cable, she took on board twenty miles of rope, which would bear a strain of thirty tons, probably the largest fishing-line used since the days of Noah..

The cable was manufactured at the rate of twenty miles a day, and as fast as delivered and found perfect, was coiled on board. And now the electricians tried their skill to outdo all that they had done before. As Captain Anderson observed, it seemed as if never had so much "brain power" been concentrated on the problem of success. The cable itself furnished the grandest subject of experiment. As every week added more than a hundred miles to its length, there was constant opportunity to try the electric current on longer distances and with new conditions. The results obtained showed the rapid and marvellous progress of electrical science. Said the London Times: "The science of making, testing, and laying cables has so much improved that an undetected fault in an insulated wire has now become literally impossible, while so much are the instruments for signalling improved, that not only can a slight fault be disregarded if necessary, but it is even easy *to work through a submarine wire with a foot of its copper conductor stripped and bare to the water.* This latter result, astonishing as it may appear, has actually been achieved for some days past with the whole Atlantic cable on board the Great Eastern. Out of a length of more than one thousand seven hundred miles, a coil has been taken from the centre, the copper conductor stripped clean of its insulation for a foot in length, and in this condition lowered over the vessel's side till it rested on the ground. Yet through this the clearest signals have been sent—so clear, indeed, as at one time to raise the question whether it would not be worth while to grapple for the first old Atlantic cable ever laid, and with these new instruments working gently through it for a year or so, at least make it pay cost."

As other things were on the same gigantic scale, by the time the big ship had her cargo and stores on board, she was well laden. Of the cable alone there were two thousand four hundred miles, coiled in three immense tanks as the year before. Of this seven hundred and forty-eight miles were a part of the cable of the last expedition. The tanks alone, with the water in them, weighed over a thousand tons; and the cable which they held, four thousand tons more; besides which she had to carry eight thousand five hundred tons of coal and five hundred tons of telegraph stores, making fourteen thousand tons, besides engines, rigging, etc., which made nearly as much more. So enormous was the burden, that it was thought prudent not to take on board all her coal before she left the Medway, especially as the channel was winding and shallow. It was therefore arranged that about a third of her coal should be taken in at Berehaven, a port on the south-west coast of Ireland. With this exception, her lading was complete.

The time for departure had been fixed for the last day of June, and so admirable had been the arrangements, and such the diligence of all concerned, that exactly at the hour of noon, she loosened from her moorings, and began to move. It was well that she had not on board her whole cargo; for as it was, she drew nearly thirty-two feet. Never had any keel pressed so deep in those waters. It required skilful handling to get her safely to the sea. Gently and softly she floated down, over bars where she almost grazed the sand, where but a few inches lifted her enormous hull above the river's bed. But at length the rising tide bears her safely over, and she is afloat in the deeper waters of the Channel. At first the sea did not give her a very gracious welcome. The wind was dead ahead, and the waves dashed furiously against her; but she kept steadily on, tossing their spray on high, as if they had struck against the rocks of Eddystone lighthouse. In four or five days she had passed down the Irish coast, and was quietly anchored in the harbor at Berehaven, where she was soon joined by the other vessels of the squadron.

The Telegraph fleet was not the same as that of the last year. The government could spare but a single ship; but the Terrible, which had accompanied the

Great Eastern on the former expedition, was still there to represent the majesty of England. The William Corry, a vessel of two thousand tons, bore the ponderous shore end, which was to be laid out thirty miles from the Irish coast, while the Albany and the Medway were ships chartered by the Company. The latter carried several hundred miles of the last year's cable, besides one of heavier proportions, ninety miles long, to be stretched across the mouth of the Gulf of St. Lawrence.

While the Great Eastern remained at Berehaven, to take in her final stores of coal, the William Corry proceeded around the coast to Valentia to lay the shore end. She arrived off the harbor on the morning of Saturday, the seventh July, and immediately began to prepare for her heavy task. This shore end was of tremendous size, weighing over eight tons to the mile. It is by far the strongest wire cable ever made, and in short lengths is stiff as an iron bar. As the year before, the cable was to be brought off on a bridge of boats reaching from the ship to the foot of the cliff. All the fishermen's boats were gathered from along the shore, while H. M. S. Racoon, which was guarding that part of the coast, sent up her boats to help, so that, as they all mustered in line, there were forty of them, making a long pontoon-bridge; and Irish boatmen with eager looks and strong hands were standing along the line, to grasp the ponderous chain. All went well, and by one o'clock the cable was landed, and its end brought up the cliff to the station. The signals were found to be perfect, and the

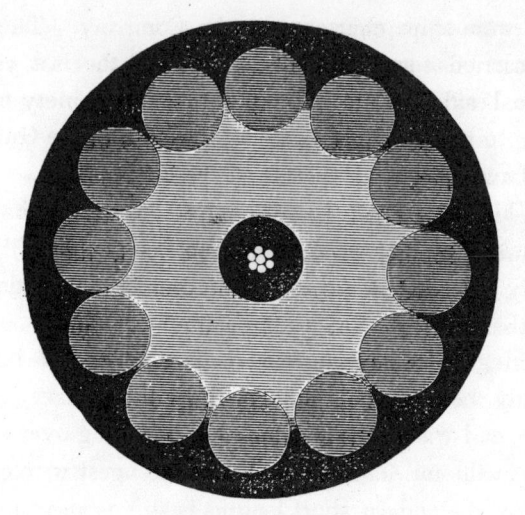

SHORE END—EXACT SIZE.

William Corry then slowly drew off to sea, unlimbering her stiff shore end, till she had cast over the whole thirty miles. At three o'clock next morning she telegraphed through the cable that her work was done, and she had buoyed the end in water a hundred fathoms deep. Describing the scene, the correspondent of the London News says:

"In its leading features it presented a striking difference to the ceremony of last year. Earnest gravity and a deep-seated determination to repress all show of the enthusiasm of which every body was full, was very manifest. The excitement was below, instead of above the surface. Speech-making, hurrahing, public congratulations, and vaunts of confidence were, as it seemed, avoided as if on purpose. There was something far more touching in the quiet and reverent solemnity of the spectators yesterday than in the slightly boisterous joviality of the peasantry last year. Nothing could prevent the scene being intensely dramatic, but the prevailing tone of the drama was serious instead of comic and triumphant. The old crones in tattered garments who cowered together, dudheen in mouth, their gaudy colored shawls tightly drawn over head and under the chin—the barefooted boys and girls, who by long practice walked over sharp and jagged rocks, which cut up boots and shoes, with perfect impunity—the men at work uncovering the trench, and winding in single file up and down the hazardous path cut by the cablemen in the otherwise inaccessible rock — the patches of bright color furnished by the red petticoats and cloaks—the ragged garments, only kept from falling to pieces by bits of string and tape—the good old parish priest, who exercises mild and gentle spiritual sway over the loving subjects of whom the ever-popular Knight of Kerry is the temporal head, looking on benignly from his car—the bright eyes, supple figures, and innocent faces of the peasant lasses, and the earnestly hopeful expression of all—made up a picture impossible to describe with justice. Add to this, the startling abruptness with which the tremendous cliffs stand flush out of the water, the alternations of bright wild flowers and patches of verdure with the most desolate barrenness, the mountain sheep indifferently cropping the short, sweet grass, and the undercurrent of consciousness of the mighty interests at stake, and few scenes will seem more important and interesting than that of yesterday."

As the ships are now ready for sea, and all who are to embark have come on board, we may look about us at the personnel of the expedition. Who are here? We recognize many old familiar faces, that we have seen in former campaigns—gallant men who have had many a sea-fight in this peaceful war. First, the eye seeks the tall form of Captain Anderson. There he is, modest and grave, of few words, but seeing every thing, watching every thing, and ruling every thing with a quiet power. And there is his second officer, Mr. Halpin, who keeps a sharp lookout after the crew, to see that every man does his duty. While he thus keeps watch of all on board, Staff Commander Moriarty, R. N., comes on deck, with instruments in hand, to look after the heavenly bodies, and reckon the ship's latitude and longitude. This is an old veteran in the service, who has been in all the expeditions, and it would be highly improper, even if it were possible, for a cable to be laid across the Atlantic without his presence and aid. And here comes Mr. Canning, the engineer, whose "deep-sea soundings," the last year, were on a scale of such magnitude, and who, if he cannot well dive deeper, means to pull stronger the next time. That slight form yonder is Professor Thomson, of Glasgow, a man who in his knowledge of the subtle element to be brought into play, and the enthusiasm he brings to its study, is the very genius of electrical science; and this is Mr. Varley, who seems to have the lightning in his fingers, and to whom the world owes some marvellous discoveries of the laws of electricity. Mr. Willoughby Smith, a worthy associate in these studies and discoveries, goes out on the ship as electrician.

And here is Mr. Glass, the managing director of the Telegraph Construction and Maintenance Company, which has undertaken by contract to manufacture this cable and lay it safely across the ocean; and Mr. Gooch, chairman of the company that owns the Great Eastern—two gentlemen to whom the Atlantic Telegraph is under the greatest obligation, since it was they who, six months before, when the project seemed in danger of being given up or postponed for years, took Mr. Field by the hand, and cheered him on to a last effort. Blessings on their hearts of oak! Mr. Gooch accompanies the ship, while Mr. Glass, keeping Mr. Varley at his side as electrician, remains on shore, there to receive reports of the daily progress of the expedition, and to issue his orders. What a post of observation was that telegraph house on the cliffs of Valentia! It was far better than the tower of Galileo, for it looked out on a broader horizon, one stretching a hundred times as far around the world. Was there ever a commander endowed with such a range of vision? What would Nelson have said, if he had had a spy-glass with which he could watch ships in action two thousand miles away, and issue his orders to a fleet on the other side of the ocean? With such a "long range," he might almost have fought the battle of the Nile from his home in England.

Standing on such a spot, and surrounded by such men, representing the capital, the science, and the skill of England, with all those gallant ships in sight, one's heart might well beat high. But there were other reflections that saddened the hour, and caused some at least to look once more on the rocks of Valentia with deep emotion. They thought of some who were not there. Brett, Mr. Field's first friend in England, was in his grave. Beyond the Atlantic, Captains Hudson and Berryman slept the sleep that knows no waking. Was it strange that they should think sadly and tenderly of the absent and the dead, and mourn that many who had toiled with them in former days, were not there now to share their triumph?

The feeling, therefore, of many on this occasion, was not one elate with pride and hope, but subdued by serious thoughts and tender memories. In harmony with this feeling, and with the great work which they were about to undertake, it was proposed that before the expedition sailed they should hold a solemn religious service.

Was there ever a fitter place or a fitter hour for prayer than here, in the presence of the great sea to which they were about to commit their lives and their precious trust? The first expedition ever sent forth had been consecrated by prayer. On that very spot, nine years before, all heads were uncovered and all forms bent low, at the solemn words of supplication; and there had the Earl of Carlisle—since gone to his honored grave—cheered them on with high religious hopes, describing the ships which were sent forth on such a mission, as "beautiful upon the waters as were the feet upon the mountains of them that publish the gospel of peace."

Full of such a spirit, officers and directors assembled

at Valentia on the day before the expedition sailed, and held a religious service. It was a scene long to be remembered. There were men of the closet and men of the field, men of science and men of action, men pale with study and men bronzed by sun and storm. All was hushed and still. Not a signal gun broke the deep silence of the hour, as, with humble hearts, they bowed together before the God and Father of all. They were about to go down to the sea in ships, and they felt their dependence on a Higher Power. Their preparations were complete. All that man could do was done. They had exhausted every resource of science and skill. The issue now remained with Him who controls the winds and waves. Therefore was it most fit that, before embarking, they should thus commit themselves to Him who alone spreadeth out the heavens, and ruleth the raging of the sea.

In all this there is something of antique stamp, something which makes us think of the sublime men of an earlier and better time; of the Pilgrim Fathers kneeling on the deck of their little ship at Leyden, as they were about to seek a refuge and a home in the forests of the New World; and of Columbus and his companions, celebrating a solemn service before their departure from Spain. And so with labor and with prayer was this great expedition prepared to sail once more from the shores of Ireland, bearing the hopes of science and of civilization — with courage and skill looking out from the bow across the stormy waters, and a religious faith, like that of Columbus, standing at the helm.

On Friday morning, the thirteenth of July, the fleet finally bade adieu to the land. Was Friday an unlucky day? Some of the sailors thought so, and would have been glad to leave a day before or after. But Columbus sailed on Friday, and discovered the New World on Friday, and so this expedition put to sea on Friday, and, as a good Providence would have it, reached land on the other side of the Atlantic on the same day of the week! As the ships disappeared below the horizon, Mr. Glass and Mr. Varley went up on their watch-tower—not to look, but to listen for the first voice from the sea. The ships bore away for the buoy where lay the end of the shore line, but the weather was thick and foggy, with frequent bursts of rain, and they could not see far on the water. For an hour or two they went sailing round and round, like sea-gulls in search of prey. At length the Medway caught sight of the buoy tossing on the waves, and, firing a signal gun, bore down straight upon it. The cable was soon hauled up from its bed, a hundred fathoms deep, and lashed to the stern of the Great Eastern; and the watchers on shore, who had been waiting with some impatience, saw the first flash, and

Varley read, "Got the shore end—all right—going to make the splice." Then all was still, and they knew that that delicate operation was going on. Quick, nimble hands tore off the covering from a foot of the shore end and of the main cable, till they came to the core; then, swiftly unwinding the copper wires, they laid them together, twining them as closely and carefully as a silken braid. Then this delicate child of the sea was wrapped in swaddling-clothes, covered up with many coatings of gutta-percha, and hempen rope, and strong iron wires, the whole bound round and round with heavy bands, and the splicing was complete. Signals are now sent through the whole cable on board the Great Eastern and back to the telegraph house at Valentia, and the whole length, two thousand four hundred and forty nautical miles, is reported perfect. And so with light hearts they bear away. It is nearly three o'clock. As they turn to the west, the following is the "order of battle": the Terrible goes ahead, standing off on the starboard bow, the Medway is on the port, and the Albany on the starboard quarter. From that hour (was it in answer to the prayers of the day before?) the voyage was a steady progress. Indeed, it was almost monotonous from its uninterrupted success. The weather was variable, alternating with sunshine and rain, fogs and squalls; but there was no heavy gale to interrupt their course, and the distance run was about the same from day to day, as the following table will show:

	Distance Run.	Cable Paid Out.
Saturday, fourteenth,	108	116
Sunday, fifteenth,	128	139
Monday, sixteenth,	115	137
Tuesday, seventeenth,	118	138
Wednesday, eighteenth,	105	125
Thursday, nineteenth,	122	129
Friday, twentieth,	117	127
Saturday, twenty-first,	122	136
Sunday, twenty-second,	123	133
Monday, twenty-third,	121	138
Tuesday, twenty-fourth,	121	135
Wednesday, twenty-fifth,	112	130
Thursday, twenty-sixth,	128	134
Friday, twenty-seventh,	112	118

This table shows the speed of the ship to have been exactly according to the "running time" before she left England. On the last voyage it was thought that she had once or twice run too fast, and thus exposed the cable to danger. It was, therefore, decided to go slowly but surely. Holding her back to this moderate pace, her average speed, from the time the splice was made till they saw land, was a little less than five nautical miles an hour, and the cable was paid out at an average of five and a half miles an hour. The total slack was less than twelve per cent, showing that the cable was laid almost in a straight line, allowing for the swells and hollows in the bottom of the sea.

As the next week drew toward its close, it was evident that they were approaching the end of their voyage. By Thursday they had passed the great depths of the Atlantic, and were off soundings. Beside their daily observations, there were many signs well known to mariners that they were near the coast. There were the sea-birds, and even the smell of the land, such as once greeted the sharp senses of Columbus, and made him sure that he was floating to some undiscovered shore. Captain Anderson had timed his departure so that he should approach the American coast at the full moon; and so, for the last two or three nights, as they drew near the Western shore, the round orb rose behind them, casting its soft light over sea and sky; and these happy men seemed like heavenly voyagers, floating gently on to a haven of rest.

In England the progress of the expedition was known from day to day, but on this side of the ocean all was uncertainty. Some had gone to Heart's Content, hoping to witness the arrival of the fleet, but not so many as the last year, for the memory of their disappointment was too fresh, and they feared the same result again. But still a faithful few were there who kept their daily watch. Two weeks have passed. It is Friday morning, the twenty-seventh of

July. They are up early and looking eastward to see the day break, when a ship is seen in the offing. She is far down on the horizon. Spy-glasses are turned toward her. She comes nearer—and look there is another and another. And now the hull of the Great Eastern looms up all-glorious in that morning sky. They are coming! Instantly all is wild excitement on shore. Boats put off to row toward the fleet. The Albany is the first to round the point and enter the Bay. The Terrible is close behind. The Medway stops an hour or two to join on the heavy shore end, while the Great Eastern, gliding calmly in as if she had done nothing remarkable, drops her anchor in front of the telegraph house, having trailed behind her a chain of two thousand miles, to bind the old world to the new.

Although the expedition reached Newfoundland on Friday, the twenty-seventh, yet, as the cable across the Gulf of St. Lawrence was broken, the news was not received in New-York till the twenty-ninth. It was early Sunday morning, before the Sabbath bells had rung their call to prayer, that the tidings came. The first announcement was brief: "Heart's Content, July 27.—We arrived here at nine o'clock this morning. All well. Thank God, the cable is laid, and is in perfect working order. Cyrus W. Field." But that was enough; and many who went up to the house of God found therein new cause for gratitude to Him whose path is in the deep waters.

Next in interest to the great event itself, was the later news it brought from Europe. While the expedition had been preparing in England, a war had broken out on the Continent of tremendous magnitude. Austria, Prussia, and Italy had rushed into the field. Armies, such as had not met since the fatal day of Leipsic, stood in battle array, and the thunder of war was echoing and reëchoing among the mountains of Bohemia.

Amid these convulsions, the fleet set sail; but it was still linked with the Old World by that cord, and received tidings from day to day. The despatch said: "We have been in constant communication with Valentia since the splice was made on the thirteenth instant, and have daily received news from Europe, which was posted up outside the telegraph office for the information of all on board of the Great Eastern, and signaled to the other ships."

This ship's log was reported here, and startled the public, just reading news a fortnight old. It told of great events; of a naval battle in the Adriatic between the Austrian and Italian fleets, and, to the surprise of all, who looked for a different result, that the Italians were defeated, and their famous iron-clads sunk! This great disaster, received with sorrow, yet showed how swift the tidings flew. But there was good news also. At the end of a column of intelligence we read, hardly believing our eyes: "*A treaty of peace has been signed between Austria and Prussia!*" This was tidings worthy to be flashed from one hemisphere to the other.

It seemed as if this new sea-nymph, rising out of the waves, was born to be the herald of peace. Eight years before, almost the first news it brought was that of peace in China. Indeed, the very first message spoken across the deep was the angels' song: "Glory to God in the highest; peace on earth, good-will toward men." And now, when again the sea "had found a tongue," its first message was Peace. Was not this "the voice of God upon the waters"?

To this best use of human speech let this new instrument be consecrated! Whatsoever goeth down into the sea, let it tend to human good. Heaven forbid that the voice of rage and anger should ever invade those tranquil depths. However men may hate each other; however they may war up here on the earth's surface, let their rage and fury, let their cursing and blasphemies, like the sound of hostile cannon, die away upon the upper air. But in that underworld, that realm of darkness and silence, where

"So lonely 'tis, that even **God**
Seems not there to be,"

let human passion never come. Peace, peace, above and below! Especially between the two kindred nations, dwelling on opposite shores of the same great sea, nations of the same blood, and speaking the same language, may this new herald of thought and speech continue to bear only messages of peace as long as the winds blow and the waters roll!